Liquid Chromatography/ Mass Spectrometry

ACS SYMPOSIUM SERIES **420**

Liquid Chromatography/ Mass Spectrometry

Applications in Agricultural, Pharmaceutical, and Environmental Chemistry

Mark A. Brown, EDITOR

California Department of Health Services

Developed from a symposium sponsored
by the Division of Agrochemicals
at the 197th National Meeting
of the American Chemical Society,
Dallas, Texas,
April 9–14, 1989

American Chemical Society, Washington, DC 1990

Library of Congress Cataloging-in-Publication Data

Liquid chromatography/mass spectrometry: applications in agricultural, pharmaceutical, and environmental chemistry
Mark A. Brown, editor.

p.   cm.—(ACS Symposium Series, ISSN 0097–6156; 420)

"Developed from a symposium sponsored by the Division of Agrochemicals at the 197th National Meeting of the American Chemical Society, Dallas, Texas, April 9–14, 1989."

Includes bibliographical references

ISBN 0–8412–1740–8

1. Liquid chromatography—Congresses. 2. Mass spectrometry—Congresses.

I. Brown, Mark A., 1953–   . II. American Chemical Society. Division of Agrochemicals. III. American Chemical Society. Meeting (197th: 1989: Dallas, Tex.). IV. Series.

QD79.C454L5536   1990
543′.0894—dc20   89–18611
                                                              CIP

The paper used in this publication meets the minimum requirements of American National Standard for Information Sciences—Permanence of Paper for Printed Library Materials, ANSI Z39.48–1984. ∞

ACS Symposium Series

M. Joan Comstock, *Series Editor*

1990 ACS Books Advisory Board

Foreword

The ACS SYMPOSIUM SERIES was founded in 1974 to provide a medium for publishing symposia quickly in book form. The format of the Series parallels that of the continuing ADVANCES IN CHEMISTRY SERIES except that, in order to save time, the papers are not typeset but are reproduced as they are submitted by the authors in camera-ready form. Papers are reviewed under the supervision of the Editors with the assistance of the Series Advisory Board and are selected to maintain the integrity of the symposia; however, verbatim reproductions of previously published papers are not accepted. Both reviews and reports of research are acceptable, because symposia may embrace both types of presentation.

Contents

INDEXES

Preface

MANY ANALYTICAL CHEMISTS HAVE RECOGNIZED the tremendous potential in coupling the resolving power of liquid chromatography with the detection specificity of mass spectrometry (LC/MS). Conventional gas chromatography (GC) methods are not adequate for many organic compounds of interest in agricultural, pharmaceutical, or environmental analyses because of limitations of high polarity and hydrophilicity, low volatility, and thermal instability. Historically available LC/MS systems (for example, direct liquid introduction and moving-belt systems, both described in the introductory chapter) have never been widely adopted. The field of LC/MS is developing so rapidly that one purpose of this volume is to provide sufficient examples of the technique to allow evaluation of its potential for other specific applications. As illustrated in this book, recent technical innovations, including the availability of the thermospray and particle beam interfaces, have made LC/MS accessible for the analysis of pesticides and pharmaceuticals and their metabolites and for environmental, forensic, polymer, dye, and process control analyses.

The first section of this book describes the application of LC/MS to the analysis of agricultural chemicals and their metabolites. Using LC/MS for residue analysis in agricultural chemistry has become routine in many laboratories. Many pesticides, such as the chlorophenoxy acid and sulfonyl urea herbicides or organophosphorus and methyl carbamate insecticides, are too polar or thermally labile for analysis via GC. The use of LC/MS for the identification of polar pesticide metabolites and conjugates, an area traditionally dominated by radiolabeled compounds, stands out as a particularly dramatic demonstration of the power of this technique.

LC/MS has had a strong influence upon pharmaceutical chemistry for the analysis of both highly polar metabolites and their precursors, as described in the second section of this book. The technique has been used for the analysis of polypeptides and a variety of intractable pharmaceuticals such as tetracycline, β-lactams, and polyether antibiotics. The structures of highly polar drug metabolites formed in vivo— including sulfate esters, glucuronides, taurine, and carnotine xenobiotic conjugates from urine, bile, and plasma as well as from in vitro metabolism experiments—have also been determined via LC/MS.

The last section describes the use of LC/MS in environmental analysis, a rapidly developing area. Current analytical methods can identify and quantify approximately 10% of the synthetic organic compounds in the average environmental matrix. The section gives examples of the use of LC/MS to measure human exposure to xenobiotics and to study detoxification processes through the analysis of target and nontarget environmental pollutants, as well as of their metabolites and conjugates and hemoglobin adducts.

I thank the authors for their time and effort in preparing chapters and the more than 40 people who reviewed the manuscripts. I also acknowledge Luis Ruzo for his invaluable help in organizing the symposium and Cheryl Shanks and Beth Ann Pratt-Dewey of the ACS Books Department for their assistance in the preparation of this book.

MARK A. BROWN
California Department of Health Services
Berkeley, CA 94704

December 5, 1989

Chapter 1

Review of the Development of Liquid Chromatography/Mass Spectrometry

Thomas Cairns and Emil G. Siegmund

Department of Health and Human Services, Food and Drug Administration, Mass Spectrometry Service Center, Los Angeles District Laboratory, 1521 West Pico Boulevard, Los Angeles, CA 90015

The inability of conventional GC/MS to adequately address the problems of analyzing molecules considered either thermally labile, of low volatility or of high polarity, has resulted in a number of different technological developments to help resolve difficult analytical situations. Over the last decade three main approaches have evolved to find practical application: the moving belt interface, direct liquid introduction and more recently in combination with supercritical fluid chromatography, and thermospray. For direct comparison of these different methods, a number of important criteria can facilitate the discussion of their relative strengths and weaknesses: eluant capacity, chromatographic resolution, sensitivity of detection, minimizing thermal degradation, choice of ionization method, limits of quantification, prolonged use, and cost factors. This tripartite array of LC/MS methods has resulted in the analytical capacity to successfully analyze those compounds normally not within the range of conventional GC/MS. The scientific literature has shown a dramatic increase in the number of LC/MS papers and reviews published that deal with analyzing polar molecules, both large and small.

The motivation behind the attention paid to the development of LC/MS was primarily brought about by the successes of commercial GC/MS systems and the strong desire on the part of the analyst to extend the range of compounds studied to include polar and thermally labile molecules. In the 1980s the rise in current awareness to the biomedical applications of mass spectrometry accelerated this curiosity initiated in the early 1970s.

Experimentally, the major problem experienced in the early development of LC interfaces was the fact that the worst scenario centered around an aqueous reversed-phase LC mobile phase at the flow rate of 1 mL/min would generate 1-4 L of gas (at 1 atm) when introduced into a MS where the vacuum was about 10^{-6} torr.   This nightmare translated into about 100 times the gas flow experienced by the early researchers into the coupling of GC with MS.   In spite of this major experimental difficulty, a myriad of solutions has been published over the last two decades which stand as testimony to the perseverance of many stubborn chemists.   No ideal universal solution to the problem has been found, but research work continues to examine alternative approaches to those in current use.

The scientific curiosity to explore the utility of mass spectrometry to compounds that could not be analyzed by conventional GC/MS was supported by the need to extend the technique into the expanding field of biochemistry. While the development of LC/MS is still undergoing rapid evolution as evidenced by the number of reviews published at regular intervals, three main technological approaches have been constructed which continue to gain popular acceptance for practical use.   These three introduction interfaces that are available commercially are the moving belt or transport interface (MBI), direct liquid introduction (DLI), and thermospray (TSP).   This review will concentrate on these three interface types that are currently in widespread use.

Many important developments have been made during the last few years but are not yet commercially available.   Specifically, there is an active interest in electrospray and continuous flow fast atom bombardment (FAB) for peptides.   Applications using supercritical fluid chromatography (SFC) in combination with MBI and DLI have been slow but continue to attract interest.   The current work in capillary zone electrophoresis (with electrospray) is extremely impressive.   The Particle Beam (PB) interface is slowly gaining strong support in pesticide residue analysis.

OPERATIONAL CHARACTERISTICS

Moving Belt Interface (MBI).   The concept of transport systems was first demonstrated by Scott et al. (1) who designed a system using a moving wire to carry the solvent/solute into the MS source via two vacuum locks where the vaporization of the solvent was accomplished.   Then vaporization of the remaining solute was carried out by passing a current through the wire. The major drawback of this early prototype to transport systems was that the efficiency of the system was a mere 1%.

In the early 1970s the Environmental Protection Agency (EPA) indicated a strong interest in LC/MS systems by contracting the Finnigan Corporation to research and develop a practical transport system.   Under the guidance of Dr. William McFadden, the sample transfer efficiency was vastly improved using a continuous polyimide Kapton belt (3 mm wide at a speed of 2.25 cm/min) permitting direct deposition of the LC eluant onto the moving belt (Fig. 1).   An infrared heater located immediately after the sample deposition zone allowed for atmospheric pressure volatilization of organic solvents.   Remaining solvent was then removed by two differentially pumped vacuum locks reducing the pressure to less than 10^{-6} torr for introduction into the MS source.   Vaporization was accomplished by a flash vaporization heater located behind the Kapton belt as it traversed the entrance to the MS source.   Early prototypes for use on quadrupole instruments were constructed such that the belt did not enter the source but merely passed an entrance leading to the source (2).   In later models, the Kapton belt was engineered to pass into the actual source for increased sensitivity (3,4).   Before recycling the belt for further sample deposition, a cleanup heater and scrubber were incorporated to avoid contamination problems.   Recent interface modifications such as a heated nebulizer for solution deposition and solvent removal have ensured chromatographic integrity of the eluted compound especially in solvent systems high in water content.   Even more recently the combination with fast atom bombardment (FAB) for the desorption and ionization of nonvolatile molecules has renewed interest in this transport type interface relative to the more modern LC/MS interfaces (3,4).   The utility of the MBI has been extensively explored by Professor D.E. Games to the analysis of a wide range of compound types and polarity illustrating the versatility of the system (5).   These applications have been responsible for the general acceptance of the MBI as a feasible LC interface.

Direct Liquid Introduction (DLI).   By far the simplest approach to LC/MS is the direct liquid introduction technique pioneered by Tal'Rose et al. (6), and Baldwin and McLafferty (7).   These interface types were designed to be compatible with the direct probe inlet systems of most commercial GC/MS systems.   A large number of such devices have been described in the literature, but the common thread of all such devices has been the dimensions of the 3 to 5 μm opening in the probe tip into the MS source (Fig. 2). The set dimensions of the orifice creates a fine liquid stream and the DLI probe is usually water cooled to prevent fluctuations in the ion current generated by the stream.   The actual flow rate is controlled by a volume restrictor based on pressure across the orifice.   Under such conditions, the normal flow rate permitted ranges from 5 to 15 μL/min.   Greater flow rates can be tolerated at the expense of expensive cryo-cooling or employment of larger capacity pumps.   More recently, however, the substitution of supercritical fluid chromatography for LC has permitted the entire effluent to enter the MS source (8) since the mobile phase (usually carbon dioxide) decompresses to a gas and is readily pumped out of the system (Fig 3).

Thermospray (TSP).   While the LC/MS interfaces described above provided partial solutions to on-line capability, the two major problems confronting such devices still remained unsolved.   These difficulties were the capability to handle flow rates of 1-2 mL/min of high polarity solvent systems and the need to analyze compounds of low volatility or poor thermal stability. The development of thermospray (TSP) mass spectrometry by Marvin Vestal (9) provided an immediate solution to both these problems.

In simplistic terms, the flow from the LC containing an electrolytic salt (usually ammonium acetate) is directed into a stainless steel hypodermic tube (0.15 mm id) which acts as a vaporizer when heated (Fig. 4). Temperature is controlled to allow the solvent stream to emerge as a fine spray at the exit of the vaporizer.   As the liquid emerges from the vaporizer tip, the aerosol droplets evaporate rapidly with a small fraction of the solution ions retaining their charge in the vapor state.   For example, with a methanol/water solvent system containing ammonium acetate the ions

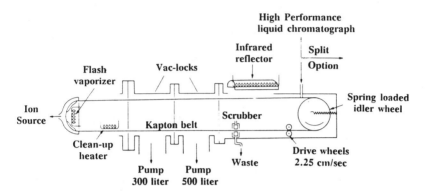

Figure 1.   Schematic diagram of a typical moving belt interface (MBI). [From reference 24 with permission]

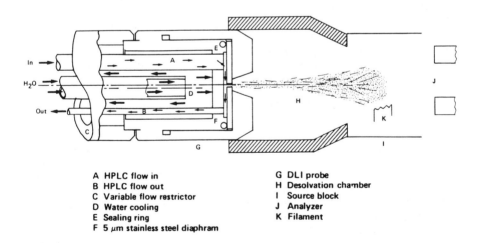

A HPLC flow in
B HPLC flow out
C Variable flow restrictor
D Water cooling
E Sealing ring
F 5 μm stainless steel diaphram

G DLI probe
H Desolvation chamber
I Source block
J Analyzer
K Filament

Figure 2.   Schematic of a direct liquid interface (DLI). [From reference 24 with permission]

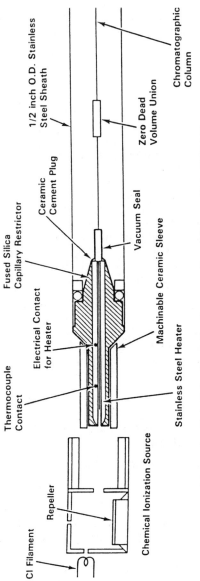

Figure 3. Detailed schematic of one version of the capillary SCF-MS interface. (Reprinted with permission from Ref. 8. Copyright 1987, Wiley.)

NH_4^+, $[H_2O]_n NH_4^+$, and $[CH_3OH]_n NH_4^+$ are the most prominent at normal operating conditions.   The lower mass ions can be favored by higher vaporizer temperatures.   Therefore, the primary source of reagent ions in TSP is the electrolyte ions in the solution.   Sample ions are then formed by interaction of the solute with the electrolyte ion, either in solution to form desolvated ions, or as the finely divided aerosol droplets evaporate. Typically, a TSP spectrum exhibits sample ions, $[M + H]^+$ and $[M + NH_4]^+$. This process represents a new method of ionization.

PERFORMANCE EVALUATION

Eluant Capacity.   Clearly TSP has the ability to handle any flow rate between 0.1 mL and 2 mL whatever the solvent system composition (10).   The only prerequisite demanded by TSP is the incorporation of a buffer such as ammonium acetate to permit ionization (11).   Recently, however, the availability of a filament for ionization has removed the absolute necessity for buffer presence.   In the case of MBI, solvent flow rates of 1.5 to 2 mL/min can be tolerated if the solvent system is non-polar.   With high water containing solvent systems, the maximum flow rate can drop to 0.1 to 0.3 mL/min via splitting.   Such splitting can effectively reduce the solute being delivered to the source by as much as one order of magnitude.   Micro column LC, however, can provide circumvention to the problem of too much water vapor entering the MS source (12).   Under DLI conditions the maximum flow rates that the pumping system can tolerate is about 5 to 50 μL/min inferring a split of 1:100.   Here again, the amount of solute being delivered to the MS source for ionization can be severely reduced.   As in the MBI the use of micro-column LC can eliminate the need for splitting (13).

Retention of Chromatographic Integrity.   Both TSP and DLI can add 1-8 seconds to the elution profile as compared to HPLC for non-volatile compounds.   The MBI, on the other hand, can reproduce the elution profiles for volatile solvent systems, but causes extreme spiking when high water containing systems are used.   This annoying feature was successfully resolved by Lankmayer et al. (14) who devised a nebulizer to spray the solvent system onto the belt rather than allow it to form droplets and causing mixing on the belt (Fig. 5).

Sensitivity of Detection.   Under TSP, a full spectrum (100 - 500 amu) can be obtained on 1 - 50 ng especially where ammonium acetate (0.1 M) has been used as buffer.   The sensitivity of detection limit observed can be extremely compound dependent.   Blakely and Vestal (15) have demonstrated impressive detection limits in the 50-100 pg range for some compounds. However, under multiple ion detection (MID) a sensitivity of about 1-50 pg can be achieved with optimized TSP conditions (16).   In the case of DLI with a split of 1:100 in the effluent, a full mass spectrum can be obtained on about 1 μg injected on column (13).   With micro column LC, a full spectrum can be obtained with 1-10 ng injected.   For the MBI a full mass spectrum can be obtained on about 25 ng injected (17).   This level can be

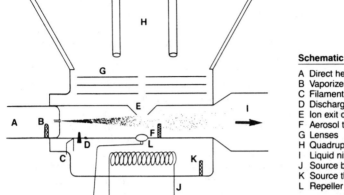

Schematic of Thermospray Interface

A Direct heated vaporizer
B Vaporizer thermocouple
C Filament*
D Discharge electrode*
E Ion exit cone
F Aerosol thermocouple
G Lenses
H Quadrupole assembly
I   Liquid nitrogen trap and forepump
J Source block heater
K Source thermocouple
L Repeller

*Dependent on model and manufacture

Figure 4.   Schematic of a thermospray LC/MS interface. [From reference 11]

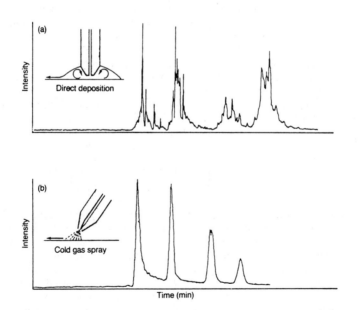

Figure 5.   Comparison of solvent deposition modes for the MBI unit in the separation of various hybrocarbons: (a) continuous flow; (b) spray deposition. [From reference 25]

improved by one order of magnitude on later models of the MBI where the Kapton belt can enter the source rather than pass outside the port leading to the MS source (18).   Under MID the level of detection is normally 100 pg entering the source.   These levels are slightly greater than achieved by conventional direct insertion probe or CI desorption mass spectrometry. With the MBI, however, the ions generated from the solvent background limits the useful lowest mass to 150 amu especially when analyzing residue samples.

Thermal Degradation.   Both TSP and DLI can analyze thermally labile and non-volatile compounds without causing thermal degradation or pyrolysis. This ability can be attributed to the very short exposures to thermal energy during their analysis.   However, in the case of the MBI the flash vapor-ization temperature of 250° C which lasts a short 0.5 s. can cause certain compounds to exhibit concentration dependent thermal degradation phenomenon (19).   In addition the MBI parts in close proximity to the source have been gold plated to reduce potential surface thermal degradation.   It should not be assumed that the various LC/MS interfaces will overcome the problems of thermal degradation.   On the contrary, the interpretation of the results should be approached so as to eliminate the presence of thermal degradation.

Choice of Ionization Method.   Both TSP and DLI can only operate to give CI spectra while the MBI can be operated under EI, CI (choice of reagent gases), NCI, FAB and SIMS.   The ability to obtained EI spectra has been stressed for library searches since under soft ionization conditions the spectra usually exhibit only protonated molecule ions and adduct ions with the reagent species and little or no fragmentation.   In this respect, the introduction of the Particle Beam LC/MS system (Fig. 6) has provided the ability to collect EI spectral data from a desolvated nebulized HPLC efflu-ent stream (20).   This emphasis placed on EI spectral data for confirmation of structure or comparison with reference standards has been successfully resolved by employing LC/MS/MS. Using protonated molecule ions as precursor ions, the resultant product spectra are highly characteristic of the com-pounds they represent.   This analytical approach has merit in that the selection of the protonated molecule ion under MID as the precursor ion permits low levels of detection as well as providing the necessary level for confirmation of presence.   Therefore, TSP, DLI and MBI have adopted tandem mass spectrometry to solve difficult structural problems (20,21).

Quantification. With the success of TSP for analyzing thermally labile compounds, attempts at quantification have indicated results with relative standard deviations in excess of the acceptable analytical limits.   This is not a surprising conclusion since TSP can exhibit a wide fluctuating vari-ation in the ionizing species and hence for the compound under study. Absolute intensity variations of 30% to 40% have been reported from day to day.   For these reasons, quantification using TSP has generally employed isotopic dilution techniques to reduce the sources of error.   Under such controlled conditions, quantification by TSP has been performed down to 100 pg with a linear dynamic range of 2 to 4 orders of magnitude (22).
    Few examples of quantification by DLI have been reported mainly due to the severe sensitivity limitation placed on the technique by splitting

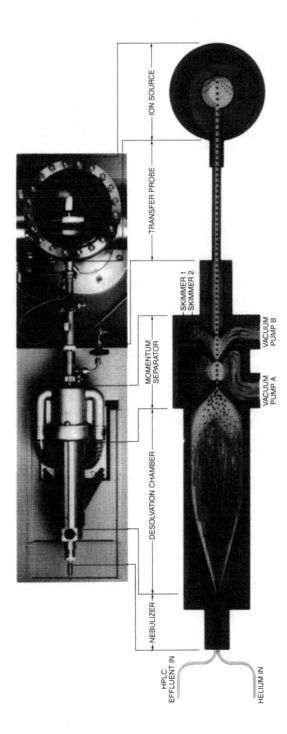

Figure 6. Cross section of the particle beam LC/MS system. [From Hewlett Packard with permission]

normal LC flow rates. However, Henion and Maylin (13) did demonstrate the determination of hydrochlorothiazide in urine by DLI-LC/MS down to 25 ng injected in column.

The utility of the MBI for quantification of residue levels of pesticides has been demonstrated by Cairns and Siegmund (23) under both EI and CI conditions. The linear dynamic range in these experiments extended from low picogram to high nanogram levels.

Ease of Operation. Since the MBI can remain installed during normal GC/MS operations, the change-over to LC/MS can be accomplished within 1 hour. The only serious malfunction is the breaking of the Kapton belt which means a shut-down and removal. Re-installation normally takes several hours.

The DLI probe can usually be fitted into the direct probe inlet and used with a CI source within a matter of minutes and operated for extended periods. Plugging of the exit orifice by a particle can reduce the operational time drastically and the need for careful preparation of solvents and cleaning LC tubing can help extend the useful operational range.

There are a number of TSP interfaces available commercially. They all usually involve the removal of the GC/MS source and replacement with a TSP source unit and a special high volume pumping system for aqueous solutions. This operation can take about one hour. One key experimental feature that must be met before installation is that the end of the vaporizer hypodermic should permit a straight flow of solvent system about 8 to 10 inches high under normal operating conditions. Recutting the end of the tubing is a technique that requires learning, but prefabricated tips from the manufacturers have eliminated this troublesome procedure. The vaporizer might require replacement about every six months depending on the extent of sediment build-up in the capillary.

Operating Factors. The MBI units range from $20,000.00 to about $35,000.00 depending on manufacturer and options. Kapton belts are about the only replacement part used on a regular basis. In the case of DLI, the price range is from $2,000.00 for a micro-LC unit to about $8,000.00 for a variable split probe with special cryo-trap. Replacement orifices cost about $45.00 each. In the case of TSP, the price range if from $24,000.00 to $35,000.00 depending on how sophisticated a unit is required. Replacement parts on an annual basis cost less than $300.00.

CONCLUSIONS

The dramatic increase in the number of publications devoted to LC/MS over the last decade is a strong indication that further progress in this field is assured. This scientific competition and exploration between the current approaches will eventually result in the development of a more universal interface. Until that time, the three major interface types will continue to be used for a ever widening variety of compound classes pushing the limits via modifications to the principal designs.

LITERATURE CITED

1. Scott, R., Scott, C., Munroe, M., and Hess, J. *J. Chromatogr.* 1974, *99*, 395.
2. McFadden, W.H. Schwartz, H., and Evans, S. *J. Chromatogr.* 1976, *122*, 389.
3. Caprioli, R.M., Fan, T., and Cottrell, J.S. *Anal. Chem.* 1986, *58*, 2949.
4. Games, D.E., Pleasance, S., Ramsey, E.D., and McDowell, M.A. *Biomed. Environ. Mass Spectrom.* 1988, *15*, 179.
5. Games, D.E. *Adv. Mass Spectrom.* 1986, *10B*, 323.
6. Tal'Rose, V., Karpov, G., Gordoetshii, I., and Skurat, V. *Russ. J. Phys. Chem.* 1968, *42*, 1658.
7. Baldwin, M. and McLafferty, F. *Org. Mass Spectrom.* 1973, *7*, 111.
8. Smith, R.D., Kalinoski, H.T., and Udseth, H.R. *Mass Spectrom. Rev.* 1987, *6*, 445.
9. Blakely, C. and Vestal, M. *Mass Spectrom. Rev.* 1983, *2*, 447.
10. Voyksner, R., and Haney, C. *Anal. Chem.* 1985, *57*, 991.
11. Voyksner, R., Bursey, J., and Pellizzari, E. *Anal. Chem.* 1984, *56*, 1507.
12. Barefoot, A.C. and Reiser, R.W. *J. Chromatogr.* 1987, *398*, 217.
13. Henion, J. and Maylin, G. *J. Anal. Toxicol.* 1980, *4*, 1985.
14. Lankmayer, E., Hayes, M., Karger, B., Vouros, P., and McGuire, J. *Int. J. Mass Spectrom. Ion Phys.* 1983, *46*, 177.
15. Blakely, C. and Vestal, M. *Anal. Chem.* 1983, *55*, 750.
16. Covey, T., Crowther, J., Dewey, E., and Henion, J. *Anal. Chem.* 1985, *57*, 474.
17. Cairns, T., Siegmund, E.G., and Doose, G.M. *Anal. Chem.* 1982, *54*, 953.
18. Games, D.E., McDowall, M.A., Lerven, K., Schafer, K.H., Dobberstein, P., and Gower, J.L. *Biomed. Mass Spectrom.* 1984, *11*, 87.
19. Cairns, T., Siegmund, E.G., Stamp, J.J., and Skelly, J.P. *Biomed. Mass Spectrom.* 1983, *10*, 203.
20. *Ion Notes*, Hewlett Packard, 1988, *3*, 2.
21. Hummel, S.V. and Yost, R.A. *Org. Mass Spectrom.* 1986, *21*, 785.
22. Liberato, D., Yergey, A., and Weintraub, S. *Biomed. Mass Spectrom.* 1986, 13, 171.
23. Cairns, T. Siegmund, E.G., and Doose, G.M. *Biomed. Mass Spectrom.* 1983, *10*, 24.
24. Voyksner, R., and Cairns, T. "Analytical Methods for Pesticides and Plant Growth Regulators", Vol. XVII, Ed. J. Sherma, Academic Press Inc., San Diego, 1989.
25. Hayes, M., Lankmayer, E., Vouros, P., Karger, B.L., and McGuire, J. *Anal. Chem.*, 1983, *55*, 1745.

RECEIVED October 6, 1989

AGROCHEMICALS: PESTICIDE METABOLISM AND DEGRADATION

Chapter 2

Techniques for Enhancing Structural Information from High-Performance Liquid Chromatography/Mass Spectrometry

Robert Voyksner[1], Terry Pack[1], Cynthia Smith[1], Harold Swaisgood[2], and David Chen[2]

[1]Research Triangle Institute, P.O. Box 12194, Research Triangle Park, NC 27709
[2]North Carolina State University, Box 7624, Raleigh, NC 27695

High performance liquid chromatography/mass spectrometry (HPLC/MS) plays an important role in characterizing nonvolatile compounds. There is a strong emphasis for obtaining structural information from HPLC/MS techniques to aid in structural elucidation. Efforts such as immobilized enzyme hydrolysis combined with thermospray HPLC/MS proved useful in characterizing endorphins. Also, the use of chemical derivatization to change proton affinity, volatility or high source temperatures to promote degradation can aid in obtaining structural information for some pharmaceuticals by thermospray HPLC/MS. Most importantly, the combination of particle beam and thermospray HPLC/MS is demonstrated to exhibit the specificity and sensitivity to characterize a wide variety of pharmaceuticals and pesticides. The choice between thermospray or particle beam HPLC/MS method depends on sample volatility, stability and proton affinity. Thermospray provided ample information for the thermally labile and nonvolatile β-lactam antibiotics. Particle beam chemical ionization provided structural information for moderately labile or semi-volatile compounds such as furosemide, spectinomycin, and some pesticides (aldicarb sulfone) and particle beam using electron ionization was useful for more stable and volatile compounds including nitrobenzamides and many pesticides.

The growing concern about residues from drugs, antibiotics, pesticides and other substances in products for human consumption has led to a need for methods to determine and confirm their presence. The thermal instability and polarity of most of these substances precludes direct analysis by gas chromatography/mass spectrometry (GC/MS). High performance liquid chromatography/mass spectrometry (HPLC/MS) methods are needed to confirm the presence of residues in food and biological matrices. HPLC/MS methods should achieve ppb sensitivity, with MS detection based upon the detection of three or more ions, and the HPLC/MS technique needs to be available as a routine technique, not only in research laboratories.

HPLC/MS employing desorption ionization techniques makes possible the analysis of many classes of nonvolatile or thermally labile compounds not successfully analyzed by other mass spectrometric techniques [1,2]. Mass spectra obtained from HPLC/MS using techniques such as thermospray [3] or fast atom bombardment (FAB) [4] consist of [M+H]$^+$ species or simple adduct ions. These ions can be advantageous for sensitivity and target compound analysis but present difficulties in structure elucidation, unknown compound identification or validating the presence of a target compound. Current research efforts are directed toward methods to enhance structural information obtained from HPLC/MS, without hindering its ability to separate and detect nonvolatile compounds, using commonly available instrumentation.

These efforts are centered around the use of techniques including enzymatic hydrolysis [5,6,7], physical or chemical degradation [8,9], and monodisperse aerosol generation interface (particle beam) for HPLC/MS [10,11] to solve specific problems. This paper discusses the implementation of several HPLC/MS methods which offer a combination of sensitivity and specificity for compounds such as peptides, pharmaceuticals, and pesticides in complex matrices.

Experimental

Sample and Solvents. Samples of peptides and pharmaceuticals were obtained from Sigma Chemical Co. (St. Louis, MO) and the pesticides were obtained from the EPA repository (Research Triangle Park, NC) and were used as received. Standard solutions (1 mg/mL) were prepared with HPLC grade methanol (Burdick and Jackson, Muskegon, MI). Dilution of the standard solutions down to 100 ng/mL were performed to determine detection limits. All solutions were stored in a freezer when not in use.

HPLC. The HPLC system consisted of two Waters Model 6000A pumps (Milford, MA) controlled by a Waters Model 720 system controller. Samples were injected with a Waters U6K injector and monitored at 254 nm with a Waters Model 440 UV detector. Separation conditions for the various classes of samples are given below:

Peptides: Separation using a BioRad RP-304 column 25 cm x 4.6 mm (Richmond, CA) using 25% isopropanol in aqueous ammonium acetate (0.1M) at a flow rate of 1 mL/min.

β-Lactams and Cephapirin: Separations were performed on a Brownlee phenyl speri 5 analytical cartridge (22 cm x 4.6 mm) (Sci Con, Winter Park, FL) using from 3-12% isopropanol, 0.5-2% acetic acid in aqueous ammonium acetate (0.1M) at a flow rate of 0.8-1.2

mL/min. Solvents and flow rates were adjusted for each sample to
maintain a clean analytical window and analysis time of under 15
minutes.

Pesticides: Separations were performed on Zorbax C_8 or ODS
column (25 cm x 4.6 mm) (DuPont Inst., Wilmington, DE) using from
50% to 70% methanol in aqueous ammonium acetate (0.1M) at a flow
rate of 0.8-1.6 mL/min.

Immobilization of Enzymes. Enzymes (carboxypeptidase A, B, and Y,
chymotrypsin, thermolysin, trypsin, and V_8-protease), obtained
from Sigma were applied directly for immobilization. About 20 mg
of each enzyme was dissolved into 0.1N phosphate buffer pH 7.0,
and placed into a 10 x 75 mm test tube with 1 g of
succinylamidopropyl glass beads. After degassing, 0.02 μmole of
1-ethyl-3-(3-dimethylaminopropyl)-carbodiimide (EDC) (Sigma
Chemical Co.) was added to the tube which then was sealed with
paraffin and rotated at 4°C overnight for simultaneous
activation/immobilization.

The immobilized enzyme was removed from the tube and rinsed
with D.I. water, followed sequentially by 2M urea solution, 2N
NaCl solution and tris buffer, pH 7.5, containing 0.2% sodium
azide and 20 mM $CaCl_2$, then stored at 4°C. About 0.18-0.2 g of
immobilized enzyme beads were used to slurry pack stainless steel
10 cm x 2.1 mm column reactors. Complete details of enzyme
preparation and assay for activity are described elsewhere [12].

Particle Beam HPLC/MS. The particle beam HPLC/MS analysis was
performed on a Hewlett-Packard 5988A (Palo Alto, CA) mass
spectrometer. The mass spectrometer was operated at 70 eV
electron energy with a source temperature ranging from 150-350°C
for acquisition of the EI spectra. The CI spectra was obtained
using methane reagent gas, at an electron energy of 100 eV, and
with a source temperature between 100-300°C. The instrument was
scanned from m/z 50-550 in 1 second. The mass spectrometer was
tuned and calibrated daily with FC-43. The particle beam
(Hewlett-Packard, Palo Alto, CA) was operated at the conditions
determined optimal from a previously reported study for the
analysis of the various drugs and antibiotics [13].

Thermospray HPLC/MS. Thermospray HPLC/MS was performed on a
Finnigan MAT 4500 quadrupole mass spectrometer (San Jose, CA).
The instrument was operated scanning from m/z 150-650 in 1 second
using positive or negative ion detection (the most sensitive mode
for the particular compound). The instrument was tuned daily and
calibrated with polypropylene glycol (MW 425 and 1000).

The thermospray interface (Vestec Inc., Houston, TX) was
operated at a vaporizer temperature of 180-195°C and a source
temperature of 250°C. The vaporizer temperature was adjusted to
optimize the solvent signal which correlated closely to optimal
conditions for the analyte [14]. Most of the compounds were
analyzed using ion evaporation ionization, although a few proved
more sensitive under discharge ionization conditions (discharge
needle at 1000 V).

Tandem MS. The thermospray HPLC/MS/MS was performed on a Finnigan MAT TSQ-46C triple quadrupole mass spectrometer interfaced to an INCOS Data System (Finnigan MAT, San Jose, CA). The triple quadrupole was operated with the first and second quadrupoles in the RF mode during HPLC/MS operation. For HPLC/MS/MS analysis, the first quadrupole selected the $[M+H]^+$ ion of the compound, while the third quadrupole was scanned over the mass range of 12-300 daltons. The second quadrupole serves as a collision chamber. Argon collision gas was added to the enclosed chamber of this quadrupole to give a pressure of 2 mtorr for collisional activation of the sample ions.

Milk Samples Preparation Procedure. A 0.5 mL aliquot of milk was diluted with 0.5 mL of a solution consisting of acetonitrile:methanol:water (40:20:40). The samples were vortexed for 10-15 seconds, placed in the microseparation system Centricon-10 (Amicon, Division of W.R. Grace Co., Danvers, MA), employing a molecular weight cutoff filter at 10,000 daltons, centrifuged for approximately 30 min at 2677 G with 45° fixed angle rotors. A 50 uL aliquot of colorless ultrafiltrate was injected for particle beam HPLC/MS for analysis.

Results and Discussion

The diversity of compounds that can be analyzed by HPLC currently preclude the use of any one HPLC/MS technique to specifically detect trace quantities. Thermospray is one of the most popular HPLC/MS techniques due to its ability to ionize nonvolatile and thermally labile compounds with minimal compromises on a HPLC separation or MS operation. Ion evaporation ionization (no filament or discharge) is suitable for numerous compounds, but often results in spectra with insufficient structural information (i.e. one ion spectra), with sensitivities varying drastically between compound classes. In cases where thermospray specificity or sensitivity is not sufficient, complementary HPLC/MS approaches need to be employed. The use of thermospray HPLC/MS and complementary techniques (i.e. immobilized enzyme bioreactors, chemical degradation and particle beam HPLC/MS) have been evaluated for the specific analysis of three major classes of compounds--peptides, pharmaceuticals, and pesticides.

Peptides. The thermospray HPLC/MS analysis of peptides result in single and multicharged protonated molecular ions. This absence of fragmentation and limited daughter ion yield for peptides above molecular weight 1800 from tandem MS dictates new analytical methodology needs to be developed. The use of immobilized enzyme hydrolysis is one method to quickly and easily provide structural information on line with HPLC/MS. The thermospray spectra shown in Figure 1 illustrate the different patterns obtained for γ-endorphin after one pass through an immobilized enzyme column. Hydrolysis by chymotrypsin or thermolysin resulted in a complex, information-rich spectra due to the greater range of peptide bonds hydrolyzed by these two endopeptidases. Also, the thermospray spectra for thermolysin and chymotrypsin hydrolysis proved quite complex due to

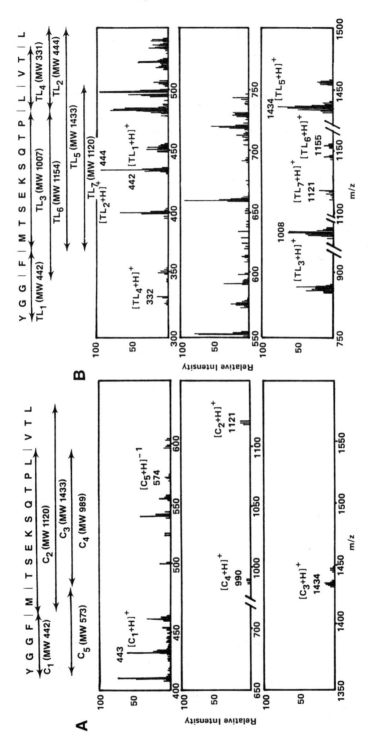

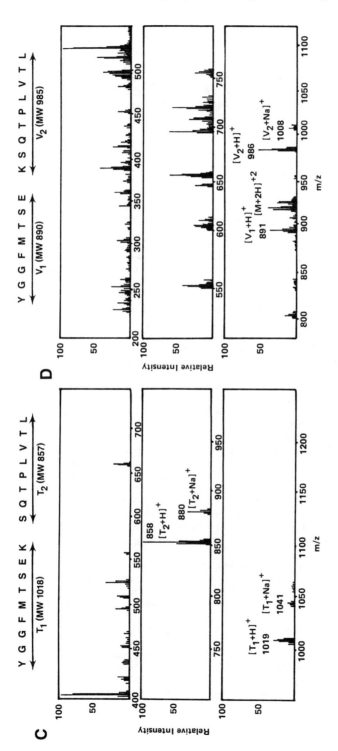

Figure 1. Thermospray mass spectra of γ-endorphin A) after one pass through a chymotrypsin 10 cm x 2.1 mm bioreactor, B) after one pass through a themolysin 10 cm x 2.1 mm bioreactor, C) after one pass through a trypsin 10 cm x 2.1 mm bioreactor, and D) after one pass through a V₈ protease 10 cm x 2.1 mm bioreactor. All spectra were acquired for 1 nmole injection of γ-endorphin in aqueous ammonium acetate (0.1 M) under identical conditions.

the incomplete hydrolysis of all peptide bonds for which it is selective.   Trypsin and V_8 protease hydrolysis spectra were more straightforward to interpret since these enzymes hydrolyze fewer peptide bonds and complete hydrolysis was observed.   In each case, enzymatic hydrolysis produced structurally confirming ions which was absent of the thermospray spectrum unhydrolyzed compound.   The thermospray spectra of the hydrolysis products from Y-endorphin exhibited $[M+H]^+$ and sometimes $[M+Na]^+$ ions.   The higher molecular weight hydrolysis products (>500 molecular weight) exhibited more intense $[M+Na]^+$ signals.

The activity of the immobilized enzyme bioreactor plays an important role in the quality of the spectra.   Immobilized enzyme activity is dependent on pH, temperature, solvents and buffers used for the hydrolysis.   Extremes in any one condition can irreversibly destroy the bioreactor activity.   Hydrolysis at conditions far from optimal can lead to no-hydrolysis or partial hydrolysis of a peptide, providing limited information on the peptide.   In order to successfully employ an immobilized enzyme column on-line with HPLC/thermospray MS, organic modifiers must be kept minimal (less than 30-50%), pH must be between 6.5 and 8.5 and buffer (ammonium acetate) concentration around 0.05-0.1M [12].

The combination of immobilized enzyme columns with HPLC/thermospray MS can be very useful in peptide identification and sequencing [6,7].   There are a number of ways of combining the immobilized enzyme column, HPLC and MS detection for peptide analysis.   Use of an endopeptidase column prior to HPLC separation and MS detection will enable separation of each hydrolysis product for identification.   Figure 2 shows the trypsin column/HPLC/thermospray MS of Y-endorphin.   The selected ion chromatograms show the retention time for each tryptic hydrolysis product T_1 and T_2.   Typically, this column configuration can only be used on purified samples since no separation or column clean-up is performed before hydrolysis.

Combination of two immobilized enzyme columns with HPLC/thermospray MS can be useful for amino acid sequencing and identification.   The use of an endopeptidase bioreactor followed by HPLC separation then an exopeptidase column and MS detection can enable sequencing of 3-5 amino acids of each endopeptidase hydrolysis product.   The trypsin, hydrolysis/HPLC/ carboxypeptidase A, B, and Y (1:1:1) hydrolysis/ thermospray MS analysis assist in the sequencing of Y-endorphin (Figure 2C,C_1).

The use of an immobilized enzyme column after the HPLC separation is useful in gaining specificity for detection of individual peptides in a complex matrix.   The separation of ∝- and Y-endorphin in plasma extract followed by tryptic hydrolysis and thermospray MS detection is shown in Figure 3.   The full scan spectra acquired contains the tryptic fragment pattern for each endorphin (similar to Figure 1C) and a weak molecular ion due to partial hydrolysis when 25% isopropanol is added to the aqueous solution, reducing enzymatic activity.   This column configuration can aid in determining whether peaks in a complex mixture are peptidic or non-peptidic.   Furthermore, a factor of 10 in sensitivity can be gained by detecting the hydrolyzed peptide

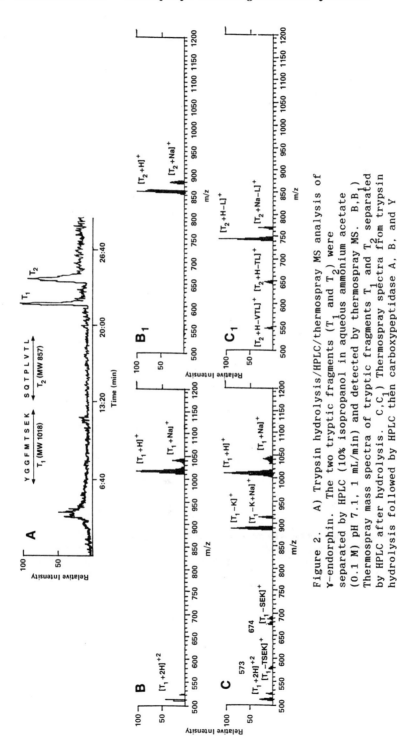

Figure 2. A) Trypsin hydrolysis/HPLC/thermospray MS analysis of γ-endorphin. The two tryptic fragments (T₁ and T₂) were separated by HPLC (10% isopropanol in aqueous ammonium acetate (0.1 M) pH 7.1, 1 mL/min) and detected by thermospray MS. B,B₁) Thermospray mass spectra of tryptic fragments T₁ and T₂ separated by HPLC after hydrolysis. C,C₁) Thermospray spectra from trypsin hydrolysis followed by HPLC then carboxypeptidase A, B, and Y hydrolysis for the partial amino acid sequencing of γ-endorphin.

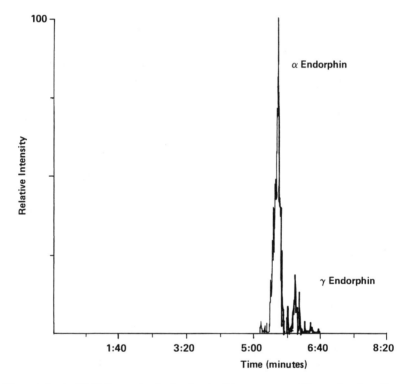

Figure 3.   HPLC/trypsin hydrolysis/thermospray MS analysis of a mixture of two endorphins in a plasma extract. The separation was performed using 25% isopropanol in aqueous ammonium acetate (0.1 M) solution at a pH of 7.5 at a flow of 1.0 mL/min.

instead of the unhydrolyzed peptide, due to superior ion yields for the lower molecular weight hydrolysis products.

Pharmaceuticals. The large variety in volatility and stability of pharmaceuticals makes it difficult for any one HPLC/MS method to address their analysis. Thermospray proved useful in analyzing a variety of β-lactam and cephapirin antibiotics. The β-lactams (5 member ring) and cephapirin (6 member ring) show several fragment ions under ion evaporation ionization. Fragmentation from ring opening and cleavage, losses of water, and thermal degradation result in the main fragment ion listed in Table 1. The β-lactams show two common fragmentations, one at m/z 160 from ring opening and cleavage, and a characteristic loss of 26, postulated to result from hydration followed by loss of CO_2.

Thermospray HPLC/MS exhibited sufficient specificity and sensitivity to confirm the presence of these β-lactams or cephapirin in dosed bovine milk. Penicillin G could be detected in bovine milk, 24 hours after dosing, at the 1 ppm level using thermospray HPLC/MS (Figure 4). Likewise cephapirin could be detected in bovine milk 48 hours after dosing. Thermospray HPLC/MS analysis of the milk extract revealed the presence of a cephapirin metabolite. The thermospray mass spectrum of the metabolite (Figure 5) indicated the molecular weight to be 381 daltons. The difference between the metabolite and cephapirin appeared to be on the 6 member ring since other fragment ions (m/z 168 and m/z 209) were present in the spectrum of cephapirin. The difference in molecular weight of 42 daltons between cephapirin and the metabolite indicate a loss of $-COCH_2$ to form desacetyl cephapirin. In vitro and in vivo studies of cephapirin in milk indicated that desacetylcephapirin predominates 16 hours after dosing. It is postulated that enzymes in milk cause the desacetylation since pasteurized milk showed no conversion of cephapirin to desacetylcephapirin.

The success for the analysis of β-lactams and cephapirin by thermospray is not shared for many pharmaceuticals. Many compounds exhibit sufficient volatility and stability that they can be analyzed by a particle beam HPLC/MS using gas phase ionization techniques. The particle beam interface opened a new avenue for HPLC/MS, permitting the acquisition of EI and CI spectra. Furthermore, the particle beam interface is simple and rugged, making it much easier to use than the moving belt HPLC/MS interface. The particle beam spectra of spectinomycin (an antibacterial) demonstrates that the acquisition of EI and CI spectra provides information for structural elucidation and for validating its identity (Figure 6). The particle beam EI and CI spectra for spectinomycin and the other compounds evaluated were independent of the HPLC and the particle beam interface conditions. The major experimental parameter evaluated that changes the qualitative appearance of the spectra was the mass spectrometer source temperature. The higher source temperatures usually result in superior particle beam ion currents but also can result in increased thermal decomposition for labile compounds. The thermospray spectrum of spectinomycin is also presented in Figure 6 for comparison to the particle beam spectra. The thermospray

Table 1. Tabular Listing of Thermospray Spectra for Several Antibiotics [1,2]

| Compound | MW | Thermospray | | |
		m/z	% RI	Identity
Cloxacillin	435	277	48	(3)
		294	40	(3)
		410	28	$[M+H+H_2O-CO_2]^+$
		436	100	$[M+H]^+$
		458	37	$[M+Na]^+$

| Cephapirin | 423 | Thermospray | | |
		m/z	% RI	Identity
		169	100	(3)+H
		209	51	(3)+H
		252	20	IU
		338	40	$[m/z\ 364+H_2O-CO_2]^+$
		364	42	$[M+H-CO_2CH_3-H]^+$
		381	31	$[M-COCH_2+H]^+$
		424	11	$[M+H]^+$

Table 1 *continued*

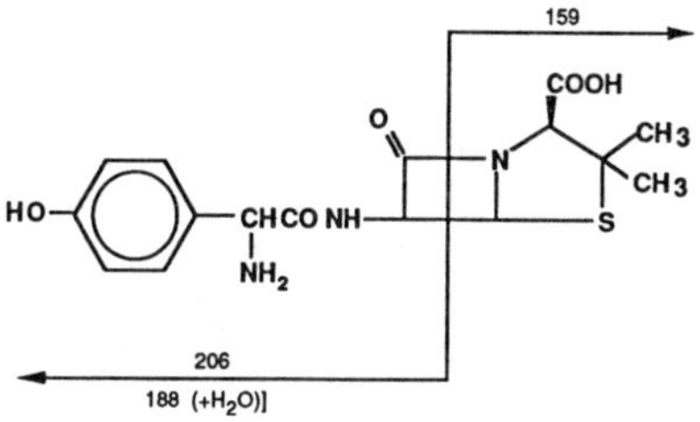

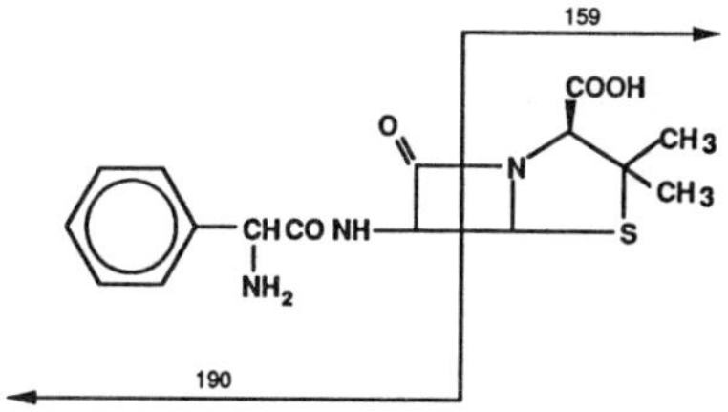

Compound	MW	Thermospray		
		m/z	% RI	Identity
Amoxicillin	365	160	100	(3) + H
		189	78	(3) + H
		207	26	(3) + H
		340	26	$[M+H+H_2O-CO_2]^+$
		366	95	$[M+H]^+$
		388	78	$[M+Na]^+$

159

COOH

CH3

CH3

O

N

S

HO— —CHCONH—

NH2

206

188 (+H2O)]

Ampicillin	349	Thermospray		
		m/z	% RI	Identity
		160	89	(3) + H
		191	72	(3) + H
		324	24	$[M+H-CO_2CH_2-H]^+$
		350	100	$[M+H]^+$
		372	52	$[M+Na]^+$

159

COOH

CH3

CH3

O

N

S

—CHCONH—

NH2

190

Penicillin G	334	Thermospray		
		m/z	% RI	Identity
		160	100	(3)+H
		176	55	(3)+H
		309	98	$[M+H+H_2O-CO_2]^+$
		335	87	$[M+H]^+$

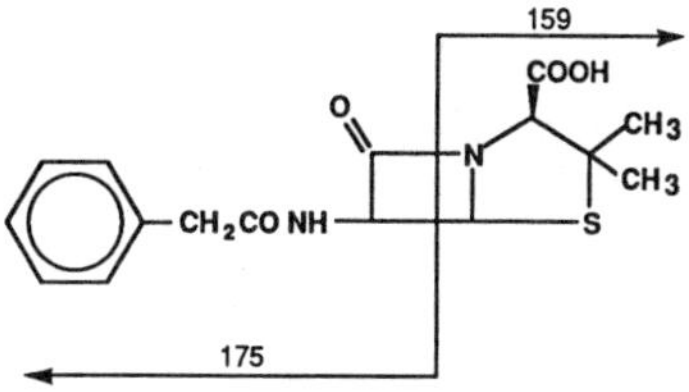

159

COOH

CH3

CH3

O

N

S

—CH2CONH—

175

IU = Identity Unknown

(1) Thermospray positive ion detection
(2) Vaporization 94 °C source 270 °C
(3) Identity shown on structure

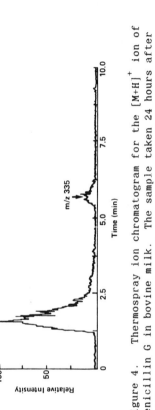

Figure 4.   Thermospray ion chromatogram for the [M+H]$^+$ ion of penicillin G in bovine milk. The sample taken 24 hours after dosing was measured to contain 1 ppm of penicillin G.

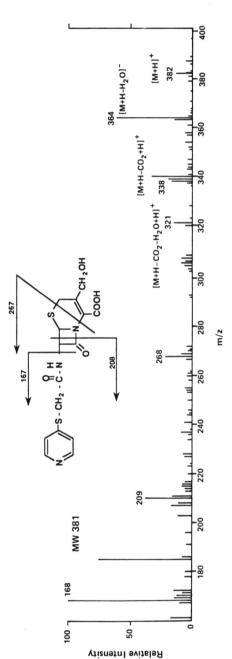

Figure 5.   Thermospray mass spectra of a metabolite of cephapirin identified as desacetylcephapirin.

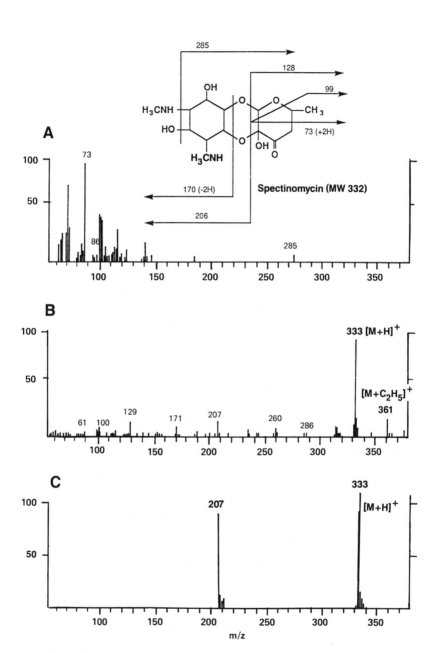

Figure 6. Mass spectra of spectinomycin: A) particle beam EI spectrum, B) particle beam methane CI spectrum, C) thermospray spectrum.

spectrum exhibited two ions, an $[M+H]^+$ ion and one fragment ion, insufficient for validating the presence of spectinomycin.

The need to acquire particle beam EI and CI spectra for compound identification and the ability of particle beam to obtain molecular weight information is compound dependent. Many compounds evaluated such as cephapirin, cloxacillin, cytidine, and penicillin G, which are thermally labile, did not yield a molecular ion and the EI spectra showed mostly low mass fragments. Methane positive ion or negative ion CI proved very useful for gaining structural information and molecular weights. The β-lactams and cephapirin were an exception, yielding thermal decomposition fragments that were also observed by thermospray MS. Other compounds including methylene blue (an antimethomoglobinemic), 2-Chlor-4-nitrobenzamide and furosemide (a diuretic) exhibited EI information by particle beam HPLC/MS (Figures 7-9). Particle beam CI of these drugs confirmed the molecular weight and showed limited high mass fragments in the case for the analysis of furosemide. The thermospray spectra for these three drugs are included for comparison, to reveal the limited fragmentation obtained using ion evaporation ionization for some classes of compounds. Particle beam HPLC/MS offered the advantage of near uniform responses for the compounds in Figures 6-9 (about 100 ng full scan EI detection limits) whereas thermospray detection limits ranged from 10 ng - 1 ug under identical MS conditions. These observations indicate the complementary nature of the particle beam and thermospray technique as well as the versatility in the ability to obtain EI, CI, and thermospray spectra to characterize a variety of pharmaceuticals.

Physical-chemical processes can be used to enhance specificity and sensitivity for some compounds analyzed by HPLC/MS. While the use of such processes can lead to solutions, their application often covers a very narrow range of compounds. Two methods whic. showed some success include: thermal degradation to improve structural information and derivatization to improve sensitivity.

Thermal degradation can take advantage of some instability found in pharmaceuticals (and some pesticides) to generate fragments. The operation of the thermospray interface at higher source and vaporization temperatures and the use of discharge CI, providing a less soft ionization than ion evaporation, can lead to fragmentation as demonstrated with furosemide (Figure 10). The thermospray spectrum of furosemide obtained under higher temperatures using discharge ionization resulted in several fragments, similar to those detected using particle beam HPLC/MS (Figure 9D). While thermal degradation can be difficult to interpret, it can provide information in the absence of particle beam or tandem MS techniques.

Chemical derivatization can be used to change the chemical characteristics of the compound to improve sensitivity or change specificity. Ionization of a compound under thermospray conditions can occur by ion evaporation or CI from reagent ions generated by a discharge or filament or from ammonium acetate present in solution [3,14]. The signal intensity for a compound is dependent on the proton affinity (PA) of the compound relative to the conjugate base for the reagent ion. For example, $[NH_4]^+$ is the main reagent ion when ammonium acetate is present, which has a PA = 206 Kcal/mole.

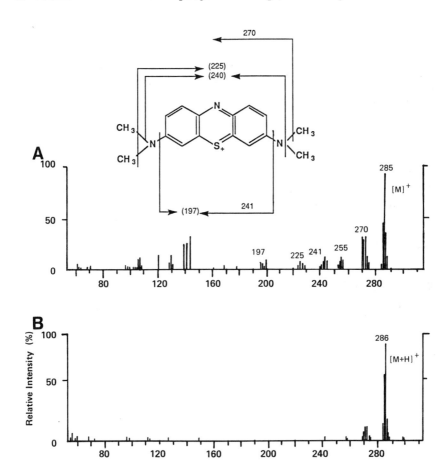

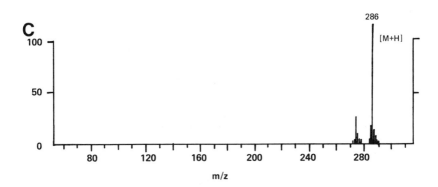

Figure 7. Mass spectra of methylene blue A) particle beam EI spectrum, B) particle beam methane CI spectrum, C) thermospray spectrum.

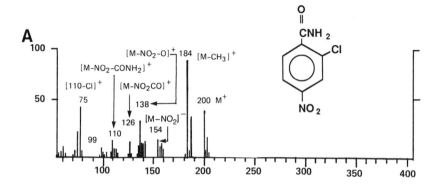

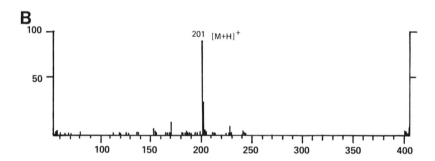

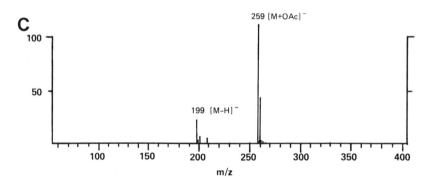

m/z

Figure 8.   Mass spectra of 2-chloro-4-nitrobenzamide: A) particle beam EI spectrum, B) particle beam methane CI spectrum, C) thermospray spectrum.

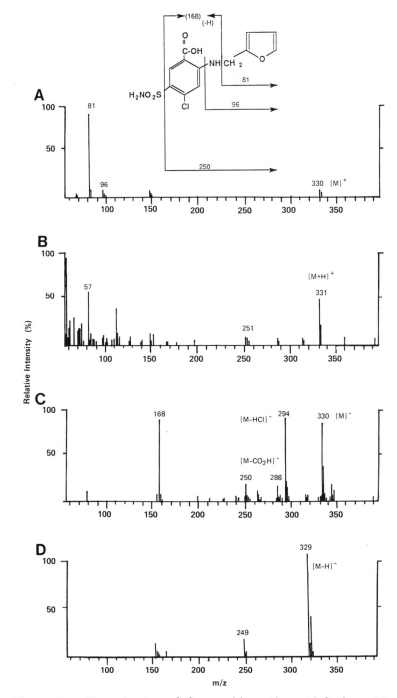

Figure 9. Mass spectra of furosemide: A) particle beam EI spectrum, B) particle beam methane CI spectrum, C) particle beam methane negative ion CI spectrum, D) thermospray spectrum.

Compounds that have PA significantly lower than ammonia (i.e.
carboxylic acids) do not undergo proton transfer reactions,
partially accounting for the diversity of sensitivities observed in
thermospray MS. Derivatization to increase the PA of a sample is
one way to increase CI sensitivity. Diethylaminoethyl (DEAE)
derivatives (reaction shown below) can change the PA of carboxylic
acids, greatly enhancing the signal level obtained by thermospray
HPLC/MS [15].

$$RCOOH + X-CH_2-CH_2N-(CH_2CH_3)_2 \rightleftharpoons RCOO-CH_2-CH_2N(CH_2CH_3)_2 + HX$$

The DEAE derivatives of prostaglandins (A_1, A_2, D_2, E_2, $F_{1\alpha}$, and
$F_{2\alpha}$) and thromboxane B_2 resulted in a thermospray MS signal level
10-30 times over the underivatized prostaglandins and thromboxane
signal level.
     Derivatization can sometimes enhance sample volatility,
improving results from HPLC/MS interface that require the sample to
be in the gas phase for ionization (i.e. particle beam and moving
belt interfaces). Limitations of the particle beam interface for
the analysis of simple organic acids such as benzoic acid can be
overcome using post column derivatization employing
trimethylanilinium hydroxide (TMAH). A 0.01M solution of TMAH
together with a desolvation temperature of 70°C could methylate a
sample of benzoic acid, improving its volatility and therefore
greatly improving the recorded signal level compared to the signal
from underivatized benzoic acid (Figure 11). The particle beam
spectrum of the TMAH derivative benzoic acid shows a molecular ion
at m/z 136 (14 daltons higher than benzoic acid) indicating the
formation of a methyl ester.

Pesticides. There are numerous references to the use of HPLC/MS
for the analysis of pesticides and herbicides [14, 16-20]. Some
major classes of pesticides and herbicides including carbamate,
triazines, organophosphorus, and phenolic acid have been analyzed
by HPLC/MS using CI or ion evaporation ionization. While these
ionization techniques often resulted in excellent sensitivity
(thermospray/MS full scan detection limits of 1-10 ng), usually
only [M+H]$^+$ and/or [M+NH$_4$]$^+$ ions were formed. This limitation can
be overcome using tandem MS [20], moving belt [17], and most
recently through the use of particle beam HPLC/MS.
     Particle beam using EI and CI ionization was useful for the
analysis of several thermally labile pesticides such as aldicarb
sulfone and fenamiphos sulfoxide, resulted in molecular ions and
structurally relevant fragment ions (Figures 12 and 13). The
thermospray spectra for these two pesticides are also presented for
comparison. The thermospray spectra consisted of only [M+H]$^+$ and
[M+NH$_4$]$^+$ ions.
     The structural information obtained from particle beam is often
similar to that obtained from tandem MS. The comparison of the
particle beam EI and the tandem MS daughter ion spectra of the
[M+H]$^+$ ion for propazine showed sufficient structural information
to identify the sample (Figure 14). The particle beam showed a
large number of simple cleavages, where as the tandem MS daughter
ion spectrum showed more rearrangements and fewer cleavages. Both

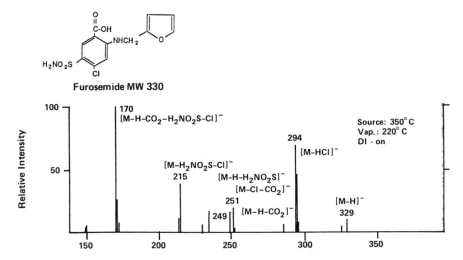

Figure 10. Thermospray spectra of furosemide acquired at a source temperature of 350°C, vaporizer temperature of 220°C with discharge ionization to promote fragmentation.

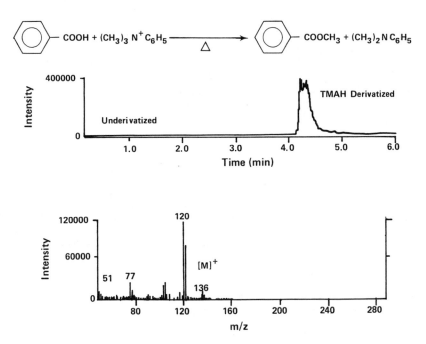

Figure 11. Post column derivatization of benzoic acid with TMAH to improve particle beam sensitivity. The particle beam EI spectrum of the methyl ester is shown on the bottom.

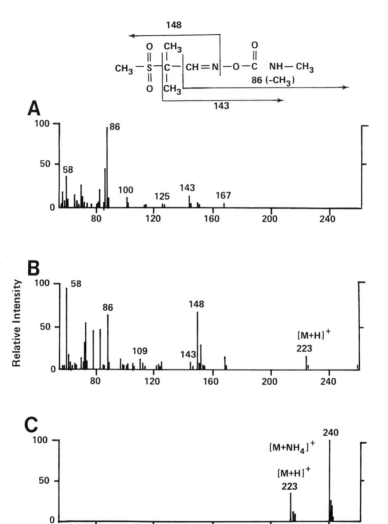

Figure 12.   Mass spectra of aldicarb sulfone: A) particle beam
EI spectrum, B) particle beam methane CI spectrum, C) thermospray
spectrum.

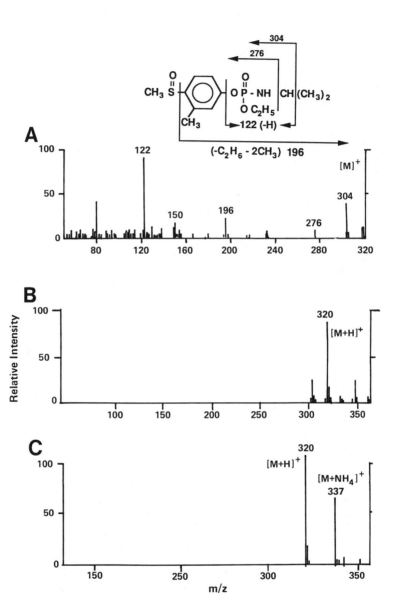

Figure 13.   Mass spectra of fenamiphos sulfoxide: A) particle beam EI spectrum, B) particle beam methane CI spectrum, C) thermospray spectrum.

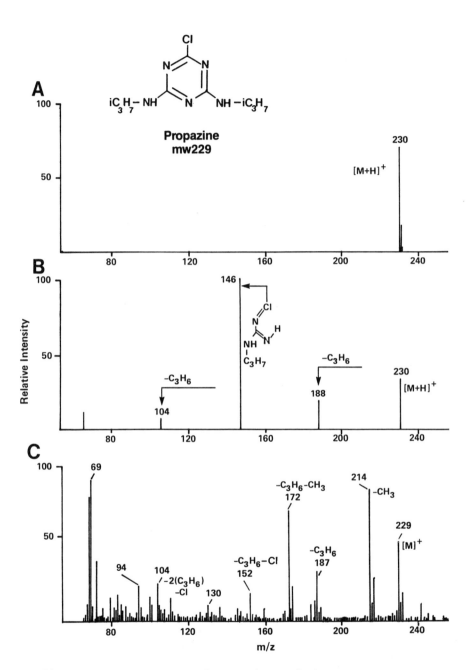

Figure 14.   Mass spectra of propazine: A) thermospray spectrum,
B) tandem MS daughter ion spectrum from the [M+H]$^+$ ion at m/z
230, C) particle beam EI spectrum.

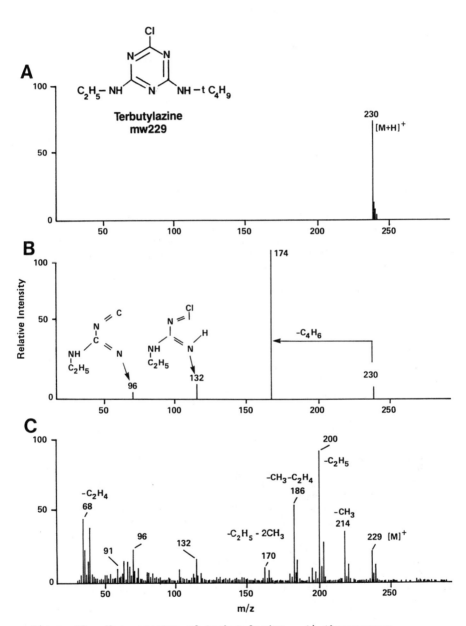

Figure 15.   Mass spectra of terbutylazine: A) thermospray
spectrum, B) tandem MS daughter ion spectrum from the [M+H]⁺ ion
at m/z 230, C) particle beam EI spectrum.

techniques could be used to identify isobaric triazines. Figures
14 and 15 show the particle EI and tandem MS spectra of propazine
and terbutazine (both have a molecular weight of 229) which could
be differentiated based on the fragments (or daughter ions)
detected. While tandem MS is valuable in understanding ion
chemistry and determining structure, more routine application and
environmental monitoring can be efficiently handled by particle
beam HPLC/MS.

Conclusions

The choice of a HPLC/MS analysis method depends greatly on the
characteristics of the sample (i.e. proton affinity, polarity,
thermal stability and volatility) as well as the structural
information and sensitivity required. The use of various
techniques including enzymatic hydrolysis and physical-chemical
reactions can assist in achieving the analysis goals for certain
compounds. Alternatively, the use of complimentary HPLC/MS
techniques such as thermospray and particle beam can be useful for
the analysis of a variety of compounds, as demonstrated in this
paper. Employing less commonly available instrumentation, such as
tandem MS, with thermospray or particle beam can prove valuable in
determining structure when other methods are unsuccessful. The
further development of existing HPLC/MS techniques and the
implantation of new HPLC/MS methods will continue to increase the
variety of compound classes that can be routinely monitored with
adequate sensitivity and specificity.

Acknowledgments

This work was partially supported by the National Institute on Drug
Abuse, Grant No. 1R01 DAC 4202-01 and by the Food and Drug
Administration, Grant No. FD-U-000244.

Literature Cited

1.   Covey, T. R.; Lee, E. D.; Bruins, A. P.; Henier, J. D. Anal.
     Chem. 1986, 58, 1451A.
2.   Vestal, M. L. Science 1984, 226, 275.
3.   Blakely, C. R.; Carmody, J. J.; Vestal, M. L. J. Am. Chem.
     Soc. 1980, 102, 5933.
4.   Barber, M.; Bordoli, R. S.; Elliott, G. J.; Sedgwick, R. D.;
     Tyler, A. N. Anal. Chem. 1982, 54, 645A.
5.   Pilosof, D.; Kim, D. -Y.; Dykes, D. F.; Vestal, M. L. Anal.
     Chem. 1984, 56, 1236.
6.   Stachowick, K.; Wilder, C.; Vestal, M. L.; Dykes, D. F. J.
     Am. Chem. Soc. 1988, 110, 1758.
7.   Kim, H. -Y.; Pilosof, D.; Dykes, D. F.; Vestal, M. L. J. Am.
     Chem. Soc. 1984, 106, 7304.
8.   Siegel, M. M.; Issnsee, R. K.; Bock, D. J. Anal. Chem. 1987,
     49, 989.
9.   Voyksner, R. D.; Bush, E. D. Biomed. Environ. Mass Spectrom.
     1987, 14, 213.

10. Winkler, P.C.; Perkins, D. D.; Williams, W. K.; Bowner, R. F.; Anal. Chem 1988, 60, 489.
11. Willoughby, R. C.; Browner, R. F. Anal. Chem. 1984, 56, 2626.
12. Voyksner, R. D.; Swaisgood, H. E.; Chen, D. Anal. Biochem., submitted, 1989.
13. Voyksner, R. D.; Knox, P.; Smith, C. S. Anal. Chem., submitted, 1989.
14. Voyksner, R. D.; Haney, C. A. Anal. Chem. 1985, 57, 991.
15. Voyksner, R. D.; Bush, E. D.; Brent, D. Biomed. Environ. Mass Spectrom. 1987, 14, 523.
16. Bellar, J. A.; Budde, W. L. Anal. Chem. 1988, 60, 2076.
17. Voyksner, R. D.; Cairns, T. Application of HPLC/MS for the Determination of Pesticides; Advanced Analytical Techniques. A Series on Analytical Methods for Pesticides and Plant Growth Regulators; Sherma, J., Ed.; Academic Press: San Diego, CA, 1989; Vol. XVII, Chapter 5, p 119.
18. Voyksner, R. D.; Bursey, J. T.; Pellizzari, E. D. Anal. Chem. 1987, 56, 1507.
19. Voyksner, R. D. Thermospray HPLC/MS for Monitoring the Environment; A Chemical Analysis Series on Application of New Mass Spectrometry Techniques in Pesticide Chemistry; Rosen, J.; J. Wiley and Sons, NY, 1987; Vol. 91, Chapter 11, p 146.
20. Voyksner, R. D.; McFadden, W. H.; Lammert, S. A. Application of Thermospray HPLC/MS/MS for the Determination of Triazine Herbicides; A Chemical Analysis Series on Application of New Mass Spectrometry Techniques in Pesticide Chemistry; Rosen, J., 1987; Vol. 91, Chapter 17, p 247.

RECEIVED November 14, 1989

Chapter 3

Confirmation of Pesticide Residues by Liquid Chromatography/Tandem Mass Spectrometry

Thomas Cairns and Emil G. Siegmund

Department of Health and Human Services, Food and Drug Administration, Mass Spectrometry Service Center, Los Angeles District Laboratory, 1521 West Pico Boulevard, Los Angeles, CA 90015

The production of a protonated molecule ion, $[MH]^+$, for a pesticide under investigation is often the principle ion produced under the soft ionization conditions determined by the various LC/MS methods. While this ion is preferred for primary identification purposes, the lack of fragment ions places the burden of proof of presence on a single ion. Product ion spectra derived from protonated molecule ions can usually provide the additional information needed to satisfy the criteria for confirmation. Cases histories discussed for organophosphorus pesticides, methylureas, and carbamates indicate a strong trend towards the increased reliance of LC/MS/MS to satisfy the criteria for confirmation.

Most applications of LC/MS to pesticide analysis involves soft ionization methods because of the operational characteristics of the various interface types employed. Only two devices can offer the option of producing spectral data under EI conditions, the Moving Belt Interface (MBI) and the more recently introduced Particle Beam. However, production of a protonated molecule ion or adduct ion provides a primary identification method based on molecular weight. The presence of a single ion is not sufficient for confirmation of presence unless high resolution measurements substantiate the molecular formula at the correct retention time. Under low resolution conditions it has been experimentally accepted that a minimum of three structurally significant ions are required for proof of presence (1). Since most applications are performed on low resolution instruments, the need for additional structural evidence for confirmation is necessary. With the availability of MS/MS instruments, the degree of specificity by soft ionization methods has been extended via precursor/product ion experiments to provide the final evidence for confirmation. In some cases, reaction product ion monitoring provides a screening method for generic classes of compounds such as organophosphorus pesticides (2,3).

CASE HISTORIES

Organophosphorus Pesticides (OP).   Because of the thermal lability of many pesticides belonging to this general class, analysis by HPLC is the method of choice for primary screening of residues.   For confirmation purposes, however, LC/MS techniques often lack sufficient structural evidence to be conclusive.   In the case of dimethoate (Fig. 1) the protonated molecule ion at m/z 230 under methane CI conditions is accompanied by two major fragment ions, m/z 88 and 199.   Under incurred residue conditions, however, the utility of the ion at m/z 88 for confirmation purposes is unrealistic, since the background spectrum generated by the MBI doers not allow ions below an m/z value of 100 to be used with confidence particularly at ng levels injected on column.   With only two potential ions for confirmation, the burden of proof of presence becomes a critical issue.   A solution to this problem can be found in the product ion spectrum generated by using the protonated molecule ion as the precursor ion (Fig. 1).   In particular, the product ion at m/z 125 is a clear indication of the presence of a dimethoxy-phosphorothionate.   The appearance of another product ion at m/z 171 is further evidence of a dithionate entity (Scheme 1).   The structural signi-ficance of these product ions derived from the protonated molecule ion without the interference of background ions (sample matrix plus interface effects) has permitted the confirmation of dimethoate beyond the minimum criteria.

In the reverse mode, product ions can be used for structural elucidation work.   In the case of etrimphos (Fig. 2), an unknown compound encountered in a pesticide residue sample, its identity could not have been inferred either by the EI or CI spectrum since both lacked fragment ions for struct-ural detective work.   However, the product ion spectrum derived from the protonated molecule yielded an indication that the compound could be a dimethoxyphosphorothionate or dithionate from the presence of the ion at m/z 125 (cf. dimethoate).   The additional presence of ions at m/z 109 and 143 in conjunction with m/z 125 was strong evidence (Fig. 2) that the compound was a dimethoxyphosphorothionate.

Additional experimental work on the MS/MS of a larger number of organo-phosphorus pesticides has revealed that generic product ions exist to characterize the phosphates, i.e. phosphorothiolates, phosphorothionates, and phosphorodithioates.   Therefore, a screening method based on certain product ions such as m/z 125 to detect the presence of dimethoxyphoro-thionates can be formulated.   This selected product ion monitoring (SPIM) approach (3) has several advantages over the conventional EI and CI spectral characterization.   First, the ability to preferentially screen without extensive sample clean-up can be an advantage over the labor intensive sample work-up thereby saving considerable time and effort.   Second, the interfering sample matrix background and the interface generated background have been eliminated.   Third, the level of sensitivity of detection for these compounds has been improved by utilizing the generic product ion approach to target those samples requiring additional analysis.   Therefore, this screening approach gives two vital criteria - the retention window containing the potential OP and its molecular weight inferred from the protonated molecule ion that yielded the product ion.

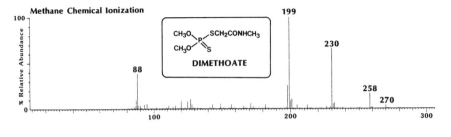

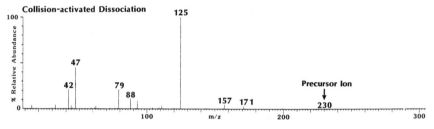

Figure 1.   Mass spectral data for dimethoate: top, methane CI; bottom, product ion spectrum using the protonated molecule ion, $[M + H]^+$, m/z 230 as the precursor ion.

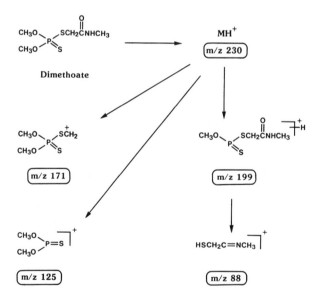

Scheme 1.   Product ion fragmentation pathway of the protonated molecule ion at m/z 230 for dimethoate using argon in the collision cell.

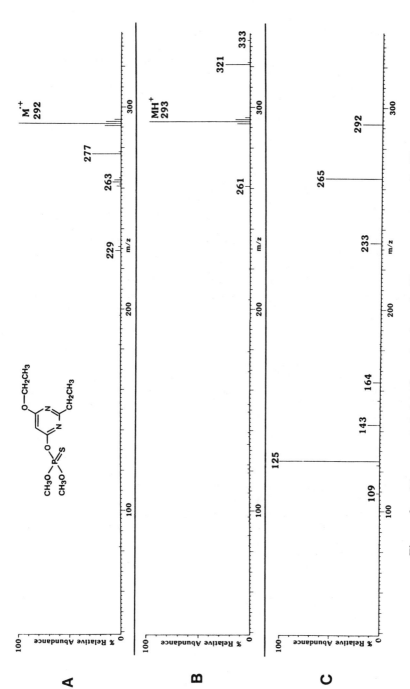

Figure 2. Mass spectral data for etrimphos: A. EI mass spectrum; B. CI mass spectrum; C, product ion spectrum using the protonated molecule ion, [M + H]$^+$, $\underline{m/z}$ 293 as the precursor ion.

Neburon [1-n-butyl-3-(3,4-dichlorophenyl)-1-methylurea].   The mass spectra obtained for neburon with the MBI under methane CI and ammonia CI are illustrated in Figure 3.   For confirmation purposes in residue samples, ion values below m/z 100 are impractical to use because of interfering background ions.   Admittedly in this case the presence of two chlorine atoms gives the added advantage of observing the appropriate chlorine ratio measurement in both the protonated molecule ion and chlorine-containing fragment ions (m/z 188 and 239 under methane CI).   These spectra were obtained using a vaporizer temperature of 170° C and a source temperature of 130° C to eliminate temperature dependence phenomenon (4).

Neburon was then examined under TSP conditions (Fig. 4).   Normally the vaporizer temperature is set for maximum sensitivity by observing the resultant intensity of the reagent gas spectrum, i.e. $[NH_4]^+$ when using ammonium acetate as buffer.   At high vaporizer temperatures, severe thermal decomposition was encountered as evidenced from the complete disappearance of the protonated molecule ion at m/z 275.   With the lowering of the temperature to 125° C, a mass spectrum was obtained displaying only two prominent fragment ions, m/z 205 and 222.   These ions were consistent with the thermal degradation of neburon to dichlorophenylisocyanate (molecular weight 187) followed by ionization of the reagent species $[NH_4]^+$ and $[NH_4 . NH_3]^+$ to give m/z 205 and 222 respectively.   Therefore, the product ion spectrum derived using the protonated molecule ion, m/z 275, was used as confirmatory proof of presence.   The ions used for confirmation were the [MH]+ and [M + H + 2] to indicate the presence of two chlorines and the product ion spectrum of $[MH]^+$ to give m/z 205 and 88.

Carbaryl [1-naphthyl-N-methylcarbamate].   The thermal instability of carbamates has necessitated their analysis by HPLC (5).   For confirmation purposes, the CI mass spectra (methane and ammonia) obtained on the MBI interface exhibit m/z 145 as the base peak corresponding to the protonated 1-naphthol moiety.   Under ammonia conditions, the mass spectrum of carbaryl exhibited a protonated molecule ion at m/z 202 together with an ammonium adduct ion at m/z 219.   However, these ions were of such low relative abundance as to preclude them from consideration as precursor ions for MS/MS experiments.   Under the circumstances it was concluded that carbaryl did not display thermal degradation and that the ion at m/z 145 was produced by fragmentation of the protonated molecule ion.

Using the protonated 1-naphthol moiety, the product ion spectrum (Fig. 5) exhibited product ions to indicate a structural disassembly of the ion atom by atom.   Similar experiments with the protonated molecule ion derived from injecting 1-naphthol did not reveal the same product ion spectrum. Therefore, for proof of presence the use of a fragment ion derived under methane CI rather than the protonated molecule ion was found to be satisfactory since it was demonstrated that 1-naphthol would not yield the same product ion spectrum.

While GC/MS methods can detect and confirm to residue levels as low as 10 pg injected, the sensitivity levels obtained by MS/MS methods are generally one order of magnitude higher (100 pg) due to loss of ions in the collision cell.

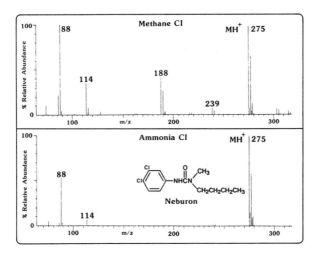

Figure 3. Mass spectral characterization of neburon via the moving belt interface: top, under methane CI conditions; bottom, under ammonia CI conditions.

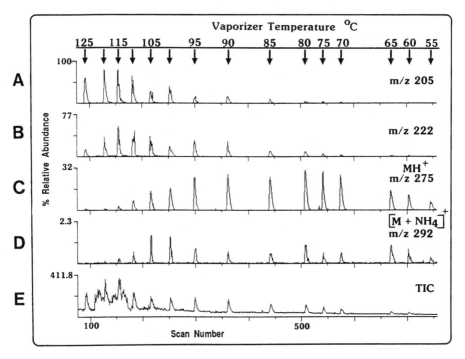

Figure 4. Multiple ion detection chromatograms for neburon via thermospray [100ng injections into the solvent system, 50% methanol-water with 0.1M ammonium acetate] at various vaporizer settings.

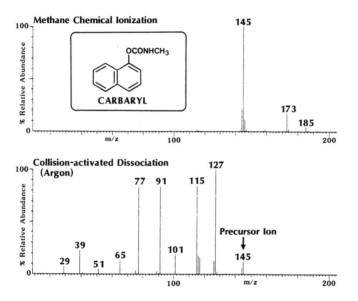

Figure 5. Mass spectral data of carbaryl via the Moving Belt Interface.

CONCLUSIONS

The real challenge in the area of trace level analysis has been the extra-polation of innovative technologies to successfully analyze those pesticides considered thermally labile or too polar for conventional analysis by GC/MS. However, this shift in emphasis to the use of the various LC/MS interfaces has now placed a new problem in the forefront of analytical strategies for confirmation of presence. The lack of sufficient fragment ions to meet the accepted criteria for confirmation has placed an increased reliance on product spectra derived mainly from the use of the protonated molecule ion as precursor ion. There can be no doubt that that experiments involving MS^N to produce a pyramid of related product ions will eventually replace the conventional attitudes of structural elucidation under CI or EI conditions.

LITERATURE CITED

1. Cairns, T.; Siegmund, E. G.; Stamp, J. J. *Mass Spectrom. Rev.* 1989, *8*, 93.
2. Hummel, S. V.; Yost, R. A. *Org. Mass Spectrom.* 1986, *21*, 785.
3. Cairns, T.; Siegmund, E. G. *J. Assoc. Off. Anal. Chem.* 1987, *70*, 858.
4. Cairns, T.; Siegmund, E. G.; Stamp, J. J. *Rapid Comm. Mass Spectrom.* 1987, *1*, 90.
5. Cairns, T.; Siegmund, E. G.; Doose, G. M.; Langham, W. S. *Bull. Environ. Contamin. Toxicol.* 1984, *32*, 310.

RECEIVED October 6, 1989

Chapter 4

Use of Solvent Adduct Ions To Confirm Structure of Selected Herbicides with Thermospray Liquid Chromatography/Mass Spectrometry

Damià Barceló

Environmental Chemistry Department, CID–CSIC, c/Jordi Girona, 18 08034 Barcelona, Spain

Conventional positive-ion and negative-ion modes (PI and NI ,respectively), the use of ammonium acetate and ammonium formate and the addition of 2% chloroacetonitrile in the liquid chromatographic eluent using filament-on thermospray LC-MS have been applied for the determination of selected herbicides. By using acetonitrile-water and 0.05 M ammonium acetate mixtures, the chlorotriazine herbicides showed $[M + (CH_3CN)H]^+$ and $[M - H]^-$ ions as base peaks in PI and NI modes, respectively. When ammonium formate and ammonium acetate were added to methanol-water eluent, the phenylurea herbicides exhibited in PI mode a relevant peak between 50-70% relative intensity corresponding to $[M + (CH_3OH)NH_4]^+$ and $[M + (CH_3COOH)NH_4]^+$, respectively, whereas the base peak was $[M + NH_4]^+$ in both cases. The formation of $[M + HCOOHNH_4]^+$ was incidently observed in PI when ammonium formate was used. With this ionizing additive and NI the chlorinated phenoxyacid herbicides exhibited $[M + HCOO]^-$ as base peak, thus allowing a complementary structural information to $[M + CH_3COO]^-$, a typical ion obtained when ammonium acetate was added as ionizing additive. The addition of 2% chloroacetonitrile to acetonitrile-water eluent showed in NI mode a $[M + Cl]^-$ ion as base peak instead of $[M + CH_3COO]^-$ or $[M + HCCO]^-$ ions in combination with other main ions which were assigned to $[M - H]^-$ and $[M + H]^-$. Applications are reported for the determination of phenylurea and chlorinated phenoxyacid herbicides in spiked water samples by using PI and NI modes, respectively.

0097–6156/90/0420–0048$06.00/0
© 1990 American Chemical Society

The on-line combination of liquid chromatography-mass spectrometry (LC-MS) plays an important role in environmental organic analysis and compared with gas chromatogrpahy-mass spectrometry (GC-MS) offers major advantages for analyzing polar pesticides and herbicides such as organophosphorus pesticides (1-5), chlorinated and phenylurea herbicides (6 - 8) , carbamates (9) and pyrethroid pesticides (10). An excellent book (11) and review articles (1,12) have been recently published reporting LC-MS applications in environmental pesticide analysis. Among the different LC-MS approaches, the thermospray (TSP) interfacing system is probably the most widely used and typically involves the use of reversed-phase columns and volatile buffers with or without a filament.
In the work described here the utility of solvent adduct ions in TSP LC-MS which consist in the use of novel additives in the chromatographic eluent, such as ammonium formate or chloroacetonitrile, will be demonstrated for confirmation of structure of a variety of herbicides including triazines, phenylurea and chlorinated phenoxyacids. Complementary adduct ion information to the conventional TSP LC-MS mode of operation will be obtained. Because TSP LC-MS involves mainly a chemical ionization process where the vaporized eluent acts as chemical ionization gas, it will be of interest to compare the different adduct ions obtained here with those using other interfacing systems such as direct liquid introduction (DLI) (13-18).

EXPERIMENTAL

Chemicals
HPLC-grade water (Farmitalia Carlo Erba, Milano, Italy), methanol (Scharlau, Barcelona, Spain) and acetonitrile (Romil, Shepshed, Leics, UK) were passed through a 0.45 μm filter (Scharlau ,Barcelona, Spain) before use. Analytical-reagent grade ammonium acetate was obtained from Panreac, (Barcelona, Spain), ammonium formate from Farmitalia Carlo Erba, (Milano, Italy), atrazine, diuron, monuron, 2,4,-D, 2,4,5-T, and silvex from Polyscience (Niles, Il, USA), linuron from Riedel-de-Haën (Seelze-Hannover, FRG) and desethylatrazine and desisopropylatrazine from Dr. Su.I. Ehrenstorfer (Augsburg, FRG)

Sample preparation
For extraction of phenylurea and chlorinated phenoxy acid herbicides in water samples, sample pre-treatment was performed by the procedures described in ref. 19 and 20, respectively. After both sample pre-treatments were finished, methanol was added to yield a final solution volume of 0.5 mL and injected into the LC-MS system.

Chromatographic conditions

Eluent delivery was provided by two model 510 high-pressure pumps coupled with automated gradient controller model 680 (Waters Chromatography Division, Millipore, MA, USA) and a model 7125 injection valve with a 20-µl loop from Rheodyne (Cotati, CA, USA). Stainless-steel columns (30 x 0.40 cm I.D.) packed from Tracer Analitica (Barcelona, Spain) with 10 µm particle diameter Spherisorb ODS-2 (Merck, Darmstadt, FRG) were used. Four different LC mobile phase compositions were tested: acetonitrile-water (50:50)   +   0.05   M   ammonium   acetate, acetonitrile-water-chloroacetonitrile (49:49:2) + 0.05 M ammonium acetate, methanol:water (50:50) + 0.1M ammonium acetate and methanol:water (50:50) + 0.1M ammonium formate at flow rates between 1-1.2 ml/min.

Mass spectrometric analysis

A Hewlett-Packard (Palo Alto, CA, USA) Model 5988A TSP LC-MS quadrupole mass spectrometer and a Hewlett-Packard Model 59970C instrument for data acquisition and processing were employed. The TSP temperatures were: stem: 100 ºC, tip: 178 ºC, vapour: 194 ºC and ion source 296 ºC with the filament on. In all the experiments the filament-on mode (ionization by an electron beam emitted from a heated filament) was used. In this mode of operation conventional positive and negative chemical ionization can be carried out by using the vaporised mobile phase as the chemical ionization reagent gas (4).

RESULTS AND DISCUSSION

Conventional PI and NI TSP LC-MS

Atrazine and two major degradation products, desethylatrazine and desisopropylatrazine have been analysed by using acetonitrile-water (50:50) and 0.05 M ammonium acetate. As regards to the chloroatrazines, it should be commented that in PI mode TSP LC-MS a common feature in the adduct ion formation has been observed, with the formation of $[M + H + CH_3CN]^+$ as base peak with the exception of desisopropylatrazine which exhibits $[M + CH_3CN]^+$ as base peak.

This fact can be attributed to its lower number of substituents and consequently lower proton affinity in comparison with the other two triazines. In contrast, it has been reported that when filament-off or thermospray ionization is employed $[M + H]^+$ is the base peak for different chloroatrazines (21,6) similarly as when DLI LC-MS was used (13). Such a difference in the relative abundance of the different adducts in the mass spectrum between filament-on and filament-off has been previously observed for other groups of pesticides (22,23). For chlorotriazines a $[M + 60]^+$ ion was the base peak using an eluent of methanol-water and ammonium

acetate (7,8). Differences in the base peak between acetonitrile-water and methanol-water mixtures can be attributed to the proton affinity values of acetic acid and acetonitrile, 797 and 798 KJ/mol, respectively, which are much higher than of methanol with a value of 773 KJ/mol (24). Table I lists the different ions obtained for atrazine by using three LC solvents in TSP LC-MS thus illustrating the dependence on adduct ion formation with the solvent used. In the Fig. 1, the PI mode TSP mass spectra of the three chloroatrazines studied are shown.
When NI mode TSP LC-MS was used for the characterization of the chloroatrazines $[M - H]^-$ was the base peak, similarly as when methanol-water mixtures in TSP for cyanazine (8) and when acetonitrile-water mixtures in DLI for atrazine (13) were employed. Sensitivities in this mode of operation were about 1 order of magnitude lower than in PI mode, in contrast to previous results with greater differences between PI and NI modes (25).

Other ionizing additives in PI TSP LC-MS
One limitation in TSP LC-MS is the need for a volatile buffer in the eluent in order to provide a soft ionization process. The most common volatile buffer used is ammonium acetate although ammonium formate has been incidently used (7,21). In this regard, a comparison between the use of both ionizing additves has been published elsewhere for the determination of a variety of herbicides(7). For phenylurea herbicides, methanol was preferred over acetonitrile as LC eluent owing to a gain (11) in sensitivity for such compounds containg aryl halide rings. The mass spectra of phenylurea herbicides monuron, diuron and linuron usually exhibit $[M + NH_4]^+$ as the base peak in the PI mode when ammonium formate or ammonium acetate are used as ionizing additives (7, 8,11), although the $[M + H]^+$ ion was the base peak for several phenylurea herbicides when filament-off TSP ionization was employed (6). The formation of $[M + H]^+$ with relative intensity values close to 20% with both ionizing additives indicates that such compounds have proton affinities below that of ammonia (858 kJ/mol) (24). When other LC-MS systems not containing ionizing additive in the eluent have been employed, such as DLI with a micro-LC (14), split (16) or moving belt (27), $[M + H]^+$ was always the base peak. In the case of using ammonium formate, the formation of adduct ions with the methanol such as $[M + (CH_3OH) NH_4]^+$ and $[M + (CH_3OH)_2 NH_4]^+$ is obtained (7). The formation of adducts with the ionizing additive, such as $[M + CH_3COOH]^+$ and $[M + (CH_3COOH) NH_4]^+$ when ammonium acetate is used (22) or $[M + (HCOOH)NH_4]^+$ in the case of ammonium formate has been also observed. The dimers $[2M + H]^+$ and $[2M + NH_4]^+$ are obtained with both ionizing additives.As an example, the different adduct ions obtained for diuron in several eluents are shown in Table I. The sensitivity was always much better in PI mode than in the NI mode of operation (about 30% lower). In contrast, other authors have found greater differences in sensitivities between both ionization modes because filament-off ionization was used (26).

TABLE I

MAIN IONS AND RELATIVE ABUNDANCES OF THREE HERBICIDES IN THERMOSPRAY LIQUID CHROMATOGRAPHY-MASS SPECTROMETRY USING DIFFERENT ELUENT MIXTURES AND FILAMENT ON
PI, and NI, positive and negative ion modes, respectively
MeCN-H_2O: acetonitrile-water ; MeOH-H_2O: methanol-water
NH_4AcO: ammonium acetate, NH_4Fo: ammonium formate
NI-Cl: 2% chloroacetonitrile

Mol wt.	Pesticide m/z ion	MeCN-H_2O (50:50) + 0.05M NH_4AcO			MeOH-H_2O (50:50) +0.1M NH_4AcO		+0.1M NH_4Fo	
		PI	NI	NI-Cl	PI	NI	PI	NI
215	Atrazine							
	214 [M − H]⁻		100	n.i.		100		100
	215 [M]⁺·	60						
	216 [M + H]⁺				50		100	
	248 [M + MeOH + H]⁺				10		50	
	257 [M + MeCN + H]⁺	100						
	275 [M + 60]⁺				100			
220	2,4-D							
	219 [M − H]⁻	n.d.	30	10	n.d.	30	n.d.	2
	255 [M + Cl]⁻		2	100				
	265 [M + Fo]⁻							100
	279 [M + AcO]⁻		100	10		100		
	311 [M + FoHFo]⁻							70
	315 [M + AcOHCl]⁻			10				
	439 [2M − H]⁻		10	7				10
232	Diuron							
	231 [M − H]⁻	n.i.	n.i.	n.i.		10		10
	233 [M + H]⁺				40		20	
	250 [M + NH₄]⁺				100		100	
	267 [M + Cl]⁻					7		
	277 [M + Fo]⁻							100
	282 [M + MeOH NH₄]⁺						80	
	291 [M + AcO]⁻					100		
	292 [M + AcOH]⁺	45						
	309 [M + AcONH₄]⁺	70						
	314 [M + (MeOH)₂NH₄]⁺						20	
	323 [M + HFoFo]⁻							70
	351 [M + AcOHAcO]⁻					80		
	465 [2M + H]⁺	50						
	482 [2M + NH₄]⁺						5	

n.i.= not investigated; n.d.= not detected.

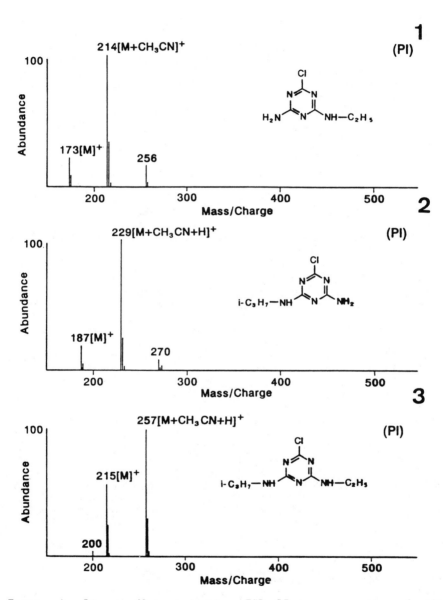

Figure 1. Direct flow injection TSP PI mass spectra of (1) desisopropyl atrazine, (2) desethyl atrazine and (3) atrazine. Carrier stream: acetonitrile-water (50:50) + 0.05 M ammonium acetate. Flow rate: 1.2 ml/min. Injection: 300 ng.

In the case of chlorotriazines differences in the base peak when using the two ionizing additives are also noticeable, with the formation of $[M + H]^+$ or $[M + 60]^{+ \cdot}$ as base peaks when either ammonium formate or ammonium acetate are used. In the NI mode, the chlorinated phenoxyacids exhibited $[M + HCOO]^-$ as base peak instead of the typical acetate adduct when ammonium formate was used instead of ammonium acetate (7). As examples, in table I the different adducts obtained in each ionizing additive for atrazine, 2,4-D and diuron are shown.

The PI TSP LC-MS traces of a water sample after the pre-treatment procedure (19) are given in Fig. 2. LC-MS using either ammonium formate or ammonium acetate as ionizing additives in the selected ion monitoring (SIM) mode is rather selective. Three different m/z values per analyte were monitored corresponding to the $[M + H]^+$, $[M + NH_4]^+$ and $[M +(CH_3COO)NH_4]^+$ and to $[M + H]^+$, $[M + NH_4]^+$ and $[M +(CH_3OH)NH_4]^+$ ions when either ammonium acetate or ammonium formate were used, respectively. No appreciable difference is observed in both chromatograms of Fig. 2, thus indicating the suitability of ammonium formate for the chromatographic separation and MS detection of such compounds.

Chloride attachment in NI TSP LC-MS

The applicability of chloride-attachment phenomenon as a way to obtain complementary structural spectral information has been studied in negative chemical ionization (NCI)-MS via direct probe (28). Parker et al (29) were the first group that demonstrated the chloride-attachment with addition of 1% chloroacetonitrile in the eluent in negative chemical ionization direct liquid introduction (NCI DLI) LC-MS using a split interface. Experiments in DLI LC-MS with total introduction from a narrow-bore LC column and a narrow-bore column have demonstrated the utility of using different amounts of chloroacetonitrile in the LC eluent for the characterization of chlorophenols, chlorinated phenoxyacids and organophosphorus pesticides (2,17). Advantages of this method were a complementary structural spectral information without a corresponding loss of sensitivity and that the chromatography was hardly influenced by the addition of this 2% of chloroacetonitrile. When the NI mode was used with acetonitrile-water mixtures containing ammonium acetate and without chloroacetonitrile in the eluent, the chlorinated phenoxyacids 2,4,-D, 2,4,5-T and silvex showed $[M + CH_3COO]^-$ ion as base peak. This base peak has been also observed previously (7) for the chlorinated phenoxy acids when methanol-water (50:50) + 0.1M ammonium acetate mixtures where employed. In contrast, other authors (11) showed $[M - H]^-$ ion as the base peak for similar experimental conditions and compounds. The formation of $[M + H]^-$ in some cases as a base or second abundant peak for 2,4,5-T and silvex needs special attention. This $[M + H]^-$ ion has been also noted in all spectra for chlorinated insecticides in negative chemical ionization MS using isobutane as reagent gas with relatively low intensities between 7-15% (30).

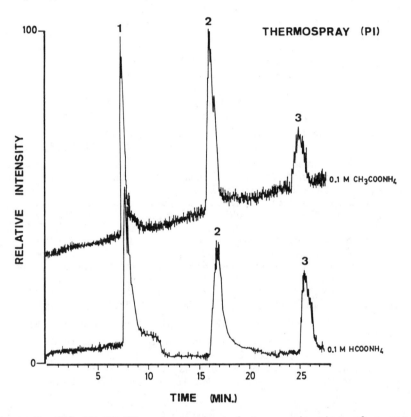

Figure 2. TSP PI LC-MS using selected ion monitoring of a water sample spiked with 3 ppm of (1) monuron, (2) diuron and (3) linuron using and LC eluent of methanol-water (50:50) + 0.1 M ammonium acetate or 0.1 M ammonium formate. Flow rate: 1 ml/min. Amount of each compound injected: 300 ng.

Such similar fragmentation behaviour between negative chemical ionization and filament-on TSP LC-MS ionization can be attributed to the virtually identical bond strengths in iso-butane and acetonitrile, 385 and 389 kJ/mol, respectively (24). These more or less parallels findings of some authors (11) who have compared TSP LC-MS filament-off ionization with ammonia chemical ionization, when using PI modes. Some chloride attachment [M + Cl]⁻ has been also observed that can be due, as in negative chemical ionization-MS (30) to a self chemical ionization process involving two molecules of parent species with the generation of the [M + Cl]⁻ ion previously observed in TSP NI MS for chlorinated containing compounds (4,7,8).

When the experiments were repeated by using 2% chloroacetonitrile in the eluent, the base peak has corresponded in most of the cases to the [M + Cl]⁻ ion, thus indicating an abundant chloride attachment. The relative abundance of the [M − H]⁻ and [M + CH₃COO]⁻ ions has decreased in comparison to the NI TSP-MS operation, without the use of chloroacetonitrile in the eluent. Another ion obtained in this case was the [M + (CH₃COOH)·Cl]⁻ ion with relative intensity values of 10 %.In Table I the different adduct ions obtained for 2,4,-D with and without the addition of chloroacetonitrile are indicated.

The mass spectra of 2,4,-D, 2,4,5-T and silvex using 0% chloroacetonitrile and 2% chloroacetonitrile in NI TSP LC-MS are shown in Fig. 3 and 4, respectively. In Fig. 5 the NI TSP LC-MS traces with 0% and 2% chloroacetonitrile in the LC eluent of a water sample spiked with 2 ppm of 2,4,-D, 2,4,5-T and silvex according to the procedure previously mentioned (20) are shown. In both cases, the MS was operated in the selected ion monitoring (SIM) mode. In the NI mode, without choroacetonitrile, the ions monitored were [M + CH₃COO]⁻ and [M − H]⁻ for 2,4,-D and 2,4,5-T and [M + CH₃COO]⁻ and [M + H]⁻ for silvex. When 2% chloroacetonitrile was added to the eluent, the ions monitored were [M + Cl]⁻ and [M − H]⁻ for 2,4,-D and silvex and [M + Cl]⁻ and [M + H]⁻ for 2,4,5-T. A more extensive study on the use of chloroacetonitrile as eluent additive in TSP LC-MS will be published elsewhere (31).

CONCLUSIONS

The use of filament-on thermospray LC-MS in environmental pesticide analysis is a valuable technique with points of similarity in the ionization process with other hyphenated systems such as direct liquid introduction (DLI) LC-MS and chemical ionization GC-MS. The relative merits of ammonium formate as ionizing additive in PI and NI modes TSP LC-MS for three different

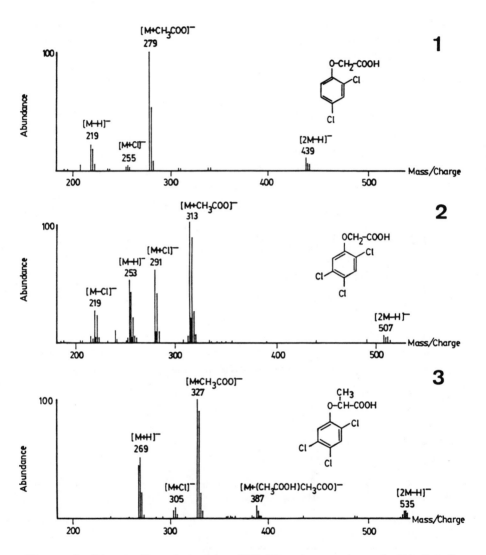

Figure 3. Direct flow injection TSP NI mass spectra of (1) 2,4-D, (2) 2,4,,5-T and (3) silvex. Carrier stream: acetonitrile-water (50:50) + 0.05 M ammonium acetate. Flow rate: 1 ml/min. Injection: 400 ng.

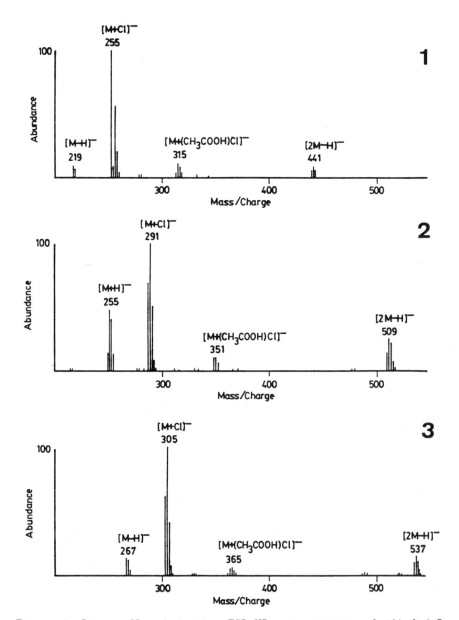

Figure 4. Direct flow injection TSP NI mass spectra of (1) 2,4-D, (2) 2,4,5-T and (3) silvex. Carrier stream: acetonitrile-water -chloracetonitrile (49:49:2) + 0.05 M ammonium acetate. Flow rate: 1 ml/min. Injection: 400 ng.

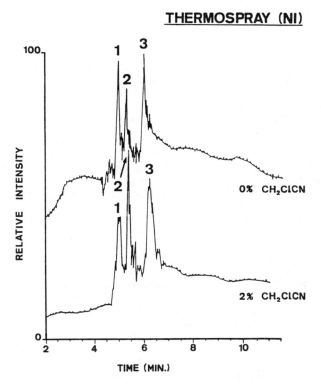

Figure 5. TSP NI LC–MS using selected ion monitoring of a water sample spiked with 2 ppm of (1) 2,4,-D, (2) 2,4,5-T and (3) silvex using and LC eluent of acetonitrile-water (50:50) + 0.05 M ammonium acetate or acetonitrile-water-chloroacetonitrile (49:49:2) + 0.05 M ammonium acetate. Flow rate: 1 ml/min. Amount of each compound injected: 200 ng.

groups of herbicides (triazines, phenylurea and chlorinated phenoxyacids) have been demonstrated. Different adduct ion formation is observed by using ammonium formate or ammonium acetate with similar sensitivities. In PI mode, for phenylurea herbicides, $[M + NH_4]^+$ was obtained as base peak in both cases whereas for chlorotriazines $[M + H]^+$ was the base peak when ammonium formate was used instead of $[M + 60]^+$ when using ammonium acetate. In NI mode, and for the chlorinated phenoxyacids, the formate or acetate adduct ions were obtained as base peaks depending on the ionizing additive employed. In the case of chlorinated phenoxyacids, the addition of chloroacetonitrile to the eluent has allowed the achievement of complementary molecular weight information with the formation of $[M + Cl]^-$ ion as base peak. As a consequence of the different solvent adducts formed, the confirmation of strucutre was feasible in both PI and NI modes of operation, either using ammonium formate and ammonium acetate for phenylurea and chlorotriazine herbicides in PI and using chloroacetonitrile in addition to both ionizing additives for phenoxyacid herbicides in NI mode.
To sum up, it can be concluded that different adduct ion formation is observed by using novel additives to the LC eluent, either in PI or NI modes, such as ammonium formate or chloroacetonitrile, respectively with no appreciable loss in senstivity compared with conventional TSP LC-MS operation. Limits of detection of 5-10 ng under full scan conditions were achieved for chlorotriazines and phenylurea herbicides in PI mode. This mode of operation was also preferred for these two groups of herbicides to NI mode, due to a gain of 3-10 fold in sensitivity. For chlorinated phenoxyacid herbicides, NI mode of operation gave a detection limits of 50 ng under full scan conditions with any of the solvents used, whereas in PI mode no signal at all was observed when 400 ng were analyzed. Future developments will include the use of other LC eluents such as cyclohexane and dichloromethane as reagent gases in TSP LC-MS in order to achieve other complementary adduct ion formation. Such eluents will be used in a post-column system and in this way the chromatographic separation will be independent from the detection allowing the use of non-volatile buffers in the eluent, similarly as previously described in DLI LC-MS (15).

ACKNOWLEDGMENTS

Financial support was provided by the C.S.I.C. and the C.A.I.C.Y.T. (Ministerio de Educación y Ciencia) and a NATO Research Grant no. 0059/88. R. Alonso (C.S.I.C.) is thanked for technical assistance.

LITERATURE CITED

1. Levsen,K. Org. Mass Spectrom., 1988, 23, 406.
2. Barceló,D.; Maris,F.A.; Geerdink,R.B.; Frei, R.W.; de Jong G.J.; Brinkman,U.A.Th. J. Chromatogr., 1987, 394 ,65.
3. Betowski, L.D.; Jones,T.L. Environ. Sci. Technol., 1988, 22, 1430.

4. Barceló,D. Biomed. Environ. Mass Spectrom. 1988, 17, 363.
5. Farran,A.; De Pablo,J; Barceló,D. J. Chromatogr. 1988, 455, 163.
6. Bellar, T.A.; Budde,W.L. Anal. Chem., 1988, 60, 2076.
7. Barceló,D. Org. Mass Spectrom., 1989, 24 219.
8. Barceló,D. Chromatographia, 1988, 25, 295.
9. Chiu,K.S.; Langenhove A.V.; Tanaka,C. Biomed. Environ. Mass Spectrom.1989, 18, 200.
10. Barceló,D. LC-GC, 1988, 6, 324.
11. Rosen, J. D.,Applications of New Mass Spectrometry Techniques in Pesticide Chemistry, Wiley, New York, 1987, pp 1-264.
12. Barceló,D. Chromatographia, 1988, 25 928.
13. Parker,C.E.; Haney, C.A.; Harvan D.J.; Hass, J.R. J. Chromatogr., 1982, 242, 77.
14. Maris,F.A.; Geerdink, R.B.; Frei, R.W; Brinkman,U.A.Th. J. Chromatogr. 1985, 323, 113
15. Apffel,J.A.; Brinkman, U.A.Th.; Frei,R.W. J. Chromatogr., 1984, 312, 153.
16. Voyksner,R.D.; Bursey,J.T.; Pellizzari, E.D. J. Chromatogr., 1984, 312, 221.
17. Geerdink,R.B.; Maris, F.A.; Frei, R.W.; de Jong, G.J.; Brinkman,U.A.Th. J. Chromatogr., 1987, 394, 51.
18. Levsen,K.; Schäfer, K.H.; Freudenthal,J. J. Chromatogr.,1983, 271, 51.
19. De Kok,A.; Opstal,M.V.; De Jong,T.; Hoogcarspel,B.; Geerdink, R.B.; Frei, R.W.; Brinkman, U.A.Th. Intern. J. Environ. Anal. Chem., 1984, 18, 101.
20. Sirons,G.J.; Chau, A.S.Y.; Smith,A.E. In Analysis of Pesticides in Water, ed. by A.S.Y. Chau and B.K. Afghan, CRC Press, Boca Raton, Fl,1982, Ch. 3.
21. Voyksner, R.D.; Haney,C.A. Anal. Chem., 1985, 57, 991.
22. Barceló, D.; Albaigés,J. J. Chromatogr., 1989, 474, 163.
23. Garteiz, D.A.; Vestal,M.L. LC-GC, 1985, 3, 335.
24. Harrison,A.G. Chemical Ionization Mass Spectrometry, CRC Press, Boca Raton, Fl, 1983, p. 33.
25. Voyksner,R.D.; Bursey, J.T.; Pellizzari,E.D. Finnigan Mat Newsletter, No. 205, San Jose, CA, 1984.
26. Voyksner,R.D.; Bursey, J.T.; Pellizzari,E.D. Anal. Chem., 1984, 56, 1507.
27. Cairns,T.; Siegmund, E.G.; Doose,G.M. Biomed. Mass Spectrom., 1983, 10, 24.
28. Dougherty,R.C.;Wander,J.D. Biomed. Mass Spectrom., 1980, 7, 401.
29. Parker,C.E.; Yamaguchi, K.; Harvan,D.J.; Smith R.W.; Hass, J.R. J. Chromatogr., 1985, 319, 273.
30. Dougherty,R.C.; Roberts, J.D.; Biros,F.J. Anal. Chem., 1975, 47, 54.
31. Vreeken,R.J.; Brinkman, U.A.Th.; de Jong, G.J.; Barceló, D. Biomed. Environ. Mass Spectrom. (in press).

RECEIVED October 6, 1989

Chapter 5

Analysis of Chlorinated Herbicides by High-Performance Liquid Chromatography/Mass Spectrometry

Tammy L. Jones[1], L. Donnelly Betowski[1], and Jehuda Yinon[2]

[1]Environmental Monitoring Systems Laboratory—Las Vegas, Office of Research and Development, U.S. Environmental Protection Agency, Las Vegas, NV 89193–3478
[2]The Weizmann Institute of Science, Rehovot, Israel

A method that uses high performance liquid chromatography/ mass spectrometry (HPLC/MS) for the analysis of chlorinated phenoxyacid herbicides is described. During method development different techniques were used to increase both the sensitivity and the specificity of thermospray HPLC/MS for chlorinated acid herbicides. These included the operation of the instrument in the negative chemical ionization (NCI) mode initiated by discharge and the use of a wire-repeller in the ion source for efficient extraction of positive ions. Single quadrupole repeller-induced and multiple quadrupole collision activated dissociation (CAD) experiments were also performed to increase the structural information of the mass spectra.

This method complements the U.S. Environmental Protection Agency's gas chromatographic (GC) SW-846 Method 8151 for the determination of chlorinated acid herbicides. Method 8151 prescribes the use of hydrolysis to ensure that these compounds are in the acid form, reaction of the acids with diazomethane to produce methyl esters, and GC or GC/MS analysis. This HPLC method eliminates the need for most sample preparation steps and allows direct analysis of sample extracts.

The use of chlorinated phenoxyacids and related compounds as herbicides originated in the 1930's with the work of Kögl who showed that indole 3-acetic acid (auxin) promotes cell elongation in plants. Many of the chlorinated phenoxyacids show auxin-like activity without being rapidly metabolized in plants. These chemicals act as herbicides by promoting uncontrolled growth in the plants. Chlorinated phenoxyacids generally have a low acute mammalian toxicity, but teratogenic effects have been observed in rodents(1) when these herbicides are administered at high dosages.

The U.S. Environmental Protection Agency has established a manual of Test Methods for Evaluating Solid Waste (SW-846) to provide a unified source of information on sampling and analysis related to compliance with the Resource Conservation and Recovery Act (RCRA). This manual provides methodology for testing representative samples of waste and other materials to be monitored. Methods 8150 and 8151 of the SW-846 manual both present protocols for determining certain chlorinated acid

[1]Current address: The Weizmann Institute of Science, Rehovot, Israel

herbicides. Both methods use gas chromatography (GC) with an electron capture detector to analyze sample extracts; Method 8150 specifies packed GC columns and Method 8151 prescribes the use of capillary GC columns. Both of these methods are labor-intensive and require the use of dangerous reagents, diazomethane or penta-fluorobenzyl bromide, to convert the acids to their corresponding methyl esters. Diazomethane is a carcinogen and can explode under certain conditions.

Because chlorophenoxy acids can be separated using high performance liquid chromatography (HPLC), the use of this technique can eliminate the need for the hydrolysis and derivatization steps required in Methods 8150 and 8151(2). The potential of thermospray high performance liquid chromatography/mass spectrometry (HPLC/MS) to measure for chlorinated phenoxyacids at the ppb level was previously shown by Voyksner(3) who operated in the negative chemical ionization (NCI) mode initiated by a filament. The NCI mode was used because detection limits when using buffer assisted ion evaporation were only 1 μg for negative ion detection and 10 μg for positive ion detection. The use of a wire-repeller(4) in the HPLC/MS interface improved the limit of detection for buffer assisted ion evaporation with positive ion analysis of a variety of compounds. This present work, therefore, has investigated thermospray introduction and various modes of ionization or dissociation for the analysis of samples containing chlorinated phenoxyacid herbicides.

Experimental

Solutions of standards were prepared from pesticides received from the Pesticide and Industrial Chemicals Repository, U.S. Environmental Protection Agency Repository (Research Triangle Park, NC).

The HPLC/MS system has been described previously(5). A Finnigan MAT TSQ 45 was interfaced to a Spectra-Physics SP8700XR gradient pump and ISCO LC-5000 syringe pump for postcolumn flow via a Vestec Thermospray system. A wire-repeller was inserted into the ion source opposite the ion exit orifice(4). Normally, the repeller was operated at 225 V in the positive ion mode and 0 V in the negative ion mode. Single quadrupole repeller-induced fragmentation spectra were obtained when this repeller was operated at 400 V and inserted 4 mm into the ion chamber.

A flow of 0.4 mL/min was used through a 15cm X 2mm ODS Hypersil 5-μm analytical column from Keystone Scientific Inc. (State College, PA). A linear solvent program of 100% water to 100% methanol in ten minutes with a 15-minute hold was used. A flow of 0.88 mL/min of 0.1 M ammonium acetate was added postcolumn to the main flow and before the LC/MS interface.

Typical operating temperatures of the thermospray interface were as follows: T(vaporizer) = 123-130°C; T(tip) = 190-210°C; T(jet) = 205-220°C; T(source) = 230-240°C.

The collision activated dissociation (CAD) experiments in the tandem mass spectrometry (MS/MS) mode were conducted at a collision energy of 20 eV with argon at a pressure of approximately 1 mTorr.

Results and Discussion

The structures of the herbicides investigated in this study are shown on Figure 1. Eight of the nine compounds are chlorinated phenoxyacids; dalapon is a chlorinated aliphatic acid. The positive ion thermospray mass spectra of these pesticides are summarized in Table I. The base peak in all cases is the $(M+NH_4)^+$ ion; for 2,4-DB the $(M+H)^+$ ion is present. No ions were detected for dalapon. Since the protonated molecule is present for 2,4-DB this compound has the highest apparent proton affinity among these herbicides.

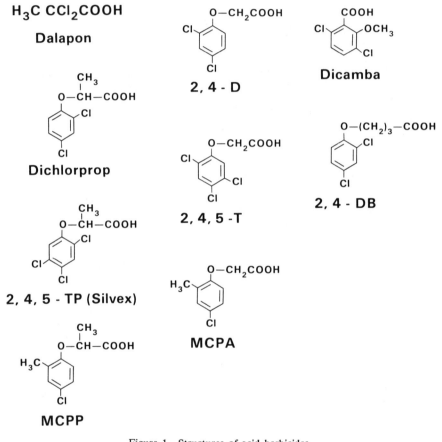

Figure 1. Structures of acid herbicides

TABLE I. Positive Ion Buffer Assisted Ion Evaporation Mass Spectra of Chlorinated Acid Herbicides

Compound	RT	MW	Mass Spectra (% Rel. Abund.)
Dicamba	2:42	220	$(M+NH_4)^+$ (100)
2,4-D	9:16	220	$(M+NH_4)^+$ (100)
MCPA	9:30	200	$(M+NH_4)^+$ (100)
Dichlorprop	10:40	234	$(M+NH_4)^+$ (100)
MCPP	10:45	214	$(M+NH_4)^+$ (100)
2,4,5-T	11:09	254	$(M+NH_4)^+$ (100)
2,4,5-TP	11:46	268	$(M+NH_4)^+$ (100)
2,4-DB	12:43	248	$(M+NH_4)^+$ (100), $(M+H)^+$ (35)

Voyksner reported 10 μg detection limits for the positive ion buffer assisted ion evaporation for similar compounds(3). Due to the increased sensitivity produced by the wire-repeller in the present work, quantities between 2 and 170 ng of the compounds in Table I could be observed in this mode.

Since two ions per compound at most are generated in the positive ion mode, little structural information is obtained from these spectra. Consequently, three options to the single quadrupole positive ion thermospray mode were tested: (1) NCI initiated by discharge; (2) daughter ion CAD in the MS/MS mode of the $(M+NH_4)^+$ ions; and (3) repeller-induced fragmentation in the positive ion single quadrupole mode.

Voyksner experienced increased sensitivity and fragmentation for the herbicides using NCI initiated by a filament(3). The results of NCI initiated by discharge from our work are summarized in Table II. The possible assignments of ions are listed for those fragments greater than 15 percent relative abundance. Along with a fairly intense $(M-H)^-$ ion, significant fragmentation occurs in these examples. Losses of Cl and $(CH_2)_xCOOH$ from the $(M-H)^-$ ion form the majority of the fragment ions. The negative ion mass spectrum of 2,4-DB is shown as an example in Figure 2. Similar ions have been observed in Voyksner's thermospray work(3) on some of these compounds. Additionally, workers using direct liquid introduction LC/MS(6,7) on chlorinated phenoxyacid herbicides have observed some of the same ions, although the present work features a greater variety of ions for each compound than any of these references. An interesting ion observed in our work and not in these experiments is the $(M+H)^-$ ion. The ion is clearly observed at m/z 255 for 2,4,5-T (Figure 3) after subtracting the ^{37}Cl isotopic contribution from the $(M-H)^-$ ion. The ion is also present in the mass spectra for 2,4,5-TP and dicamba (Table II). The $(M+H)^-$ ion has been observed previously. This ion has been reported in the isobutane negative chemical ionization spectra of o,p'-DDT(8) and two nitrated explosives(9). Barceló has also observed this ion for silvex(10). This ion apparently is formed from ion-molecule reactions with M⁻ or $(M-H)^-$. The limits of detection (LOD) ranged from 3 to 114 ng under NCI initiated by discharge and are summarized in Table III together with the values under positive ion buffer assisted ion evaporation.

TABLE II.   Negative Ion Mass Spectra of Chlorinated Acid
Herbicides (Discharge On)

Compound	MW	Mass Spectra (% Relative Abundance)
Dalapon	142	$(M+OAc)^-(25)$, $(M-H)^-(100)$
Dicamba	220	$(M+H)^-(33)$, $M^-(39)$, $(M-H)^-(33)$ $(M-Cl)^-(60)$, $(M-HCl)^-(100)$, $(M-CH_2Cl)^-(31)$, $(M-H_2Cl_2)^-(45)$
2,4-D	220	$M^-(60)$, $(M-H)^-(100)$, $(M-Cl)^-(35)$, $(M-HCl)^-(60)$, $(M-HCOOH)^-(43)$, $(M-CH_2COOH)^-(20)$
MCPA	200	$(M+HCO_2)^-(10)$, $(M-H)^-(100)$
Dichlorprop	234	$M^-(17)$, $(M-H)^-(100)$, $(M-Cl)^-(20)$, $(M-HCl)^-(30)$, $(M-CH(CH_3)COOH)^-(22)$
MCPP	214	$(M+HCO_2)^-(7)$, $(M-H)^-(100)$
2,4,5-T	254	$(M+H)^-(31)$, $M^-(27)$, $(M-H)^-(27)$, $(M-Cl)^-(22)$, $(M-HCl)^-(100)$, $(M-HCOOH)^-(18)$, $(M+H-CH_2COO)^-(33)$, $(M-CH_2COOH)^-(15)$, $(M-Cl-COOH)^-(38)$, $(M-Cl-CHCOOH)^-(23)$
2,4,5-TP	268	$(M+H)^-(73)$, $M^-(13)$, $(M-H)^-(36)$, $(M-Cl)^-(28)$, $(M-HCl)^-(83)$, $(M+H-Cl_2)^-(58)$, $(M-HCl_2)^-(100)$, $(M-CH(CH_3)COOH)^-(28)$, $(M-Cl-C(CH_3)COOH)^-(35)$, $(M-H_2Cl_3)^-(31)$
2,4-DB	248	$(M+HCO_2)^-(20)$, $M^-(10)$, $(M-H)^-(100)$, $(M-(CH_2)_3COOH)^-(77)$

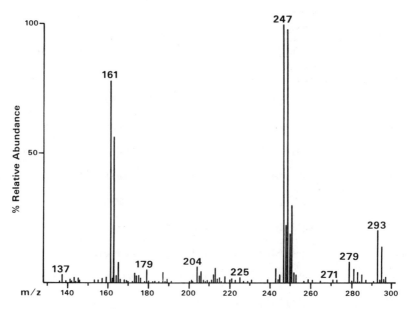

Figure 2. Negative ion spectrum of 2,4-DB (discharge on)

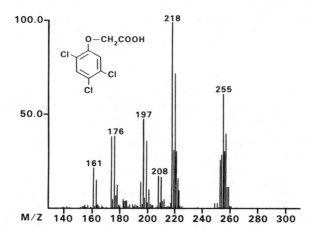

Figure 3. Negative ion spectrum of 2,4,5-T (discharge on)

TABLE III. Limits of Detection in the Positive and Negative Ion
Modes for the Chlorinated Herbicides

Compound	Positive Mode Quantification Ion	LOD (ng)	Negative Mode Quantification Ion	LOD (ng)
Dalapon	Not detected		141(M-l)⁻	11
Dicamba	238(M+NH₄)⁺	13	184(M-HCl)⁻	3.0
2,4-D	238(M+NH₄)⁺	2.9	219(M-l)⁻	50
MCPA	218(M+NH₄)⁺	120	199(M-l)⁻	28
Dichlorprop	252(M+NH₄)⁺	2.7	235(M-l)⁻	25
MCPP	232(M+NH₄)⁺	5.0	213(M-l)⁻	12
2,4,5-T	272(M+NH₄)⁺	170	218(M-HCl)⁻	6.5
Silvex	286(M+NH₄)⁺	160	269(M+1)⁻	43
2,4-DB	266(M+NH₄)⁺	3.4	247(M-1)⁻	110

The use of a wire-repeller at a nonzero voltage was investigated for NCI initiated by discharge. Fink and Freas(11) reported enhancement in both positive and negative modes of buffer assisted ion evaporation. We found a more complex dependence of the negative ion intensity (discharge on) with repeller voltage. Initially, a decrease in the analyte signal was observed followed by an increase to a maximum at approximately 100 V followed by another decrease in signal. Since the enhancement at 100 V was only a factor of two or less, the limits of detection measurements for the negative ions were performed at a repeller setting of 0 V. With the discharge off, the negative ions observed under buffer assisted ion evaporation showed a small initial enhancement followed by a decrease in the signal which is in agreement with the findings of Fink and Freas.

The information generated in the MS/MS mode of operation of the triple quadrupole mass spectrometer for these chlorinated herbicides is typified by the spectrum shown in Figure 4. This figure represents the daughter ion CAD spectrum of the m/z 266 ion from 2,4-DB. The $(M+NH_4)^+$ ion at m/z 266 was generated in the buffer assisted ion evaporation of 2,4-DB. Losses of OH (or NH_3), Cl (or NH_3 and H_2O), and $(CH_2)_xCOOH \cdot NH_4$ from the ammoniated species comprise the majority of the ions in the spectrum; $(CH_2)_3COOH^+$ is the base peak. Since the m/z 266 ion has only contributions from ^{35}Cl, the CAD spectrum is not marked by the chlorine isotopic pattern that these herbicides usually show (see Figure 2). Consequently, a simpler spectrum results.

Attempts to increase the voltage on the wire-repeller above the value that maximized the sensitivities for these compounds resulted in additional fragmentation of the chlorinated herbicides. The wire-repeller was also moved closer to the sampling

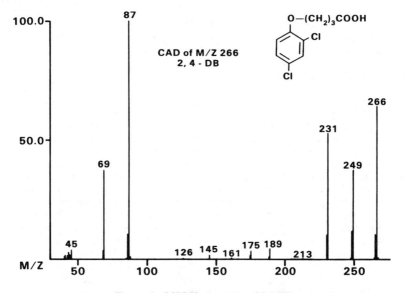

Figure 4. MS/MS spectrum of 2,4-DB

orifice. Bencsath and Field reported similar effects in a system with a retarding electrode downstream of the thermospray gas stream(12). McFadden and Lammert also reported seeing fragmentation with an increased voltage on a repeller in a thermospray ion source(13). A loss of sensitivity resulted in our work from operating off this maximum value and from fragmentation, thereby spreading the total ionization over more channels. Figures 5 and 6 show the single quadrupole spectra of 2,4-D and 2,4-DB, respectively, generated with the wire-repeller at 400 V. This spectrum for 2,4-D shows similarities with its electron impact spectrum(14), as m/z 162 $(C_6H_3Cl_2OH)^+$ is the base peak in both cases. Also, comparing the MS/MS CAD spectrum of 2,4-DB (Figure 4) with this repeller-induced fragmentation spectrum (Figure 6) is instructive. It appears that the m/z 231 ion, which is a major peak in both spectra, is due to the loss of H_2O from the $(M+H)^+$ ion, rather than loss of Cl from the $(M+NH_4)^+$ ion. The repeller-induced fragmentation spectrum shows the same isotopic pattern for both the $(M+1)^+$ ion and the $(M+1-18)^+$ ion. The m/z 162 peak is very intense in the repeller-induced fragmentation spectrum while this peak does not appear in the MS/MS CAD spectrum. The $(CH_2)_3COOH^+$ ion is the base peak in both spectra.

Table IV presents the repeller-induced fragmentation mass spectra for eight acid herbicides. Again, possible ions are assigned for those peaks greater than 15 percent relative abundance. The limits of detection for the herbicides in this mode were approximately 1-5 μg, while similar limits under MS/MS CAD conditions were approximately 250-1000 ng.

Conclusions

Two important features that should characterize a mass spectral-based method for the analysis of environmental samples are sensitivity and specificity. Sensitivity may be compound dependent, but it can also change under different modes of ionization. Since the phenoxyacid herbicides that were tested in the present study are chlorinated or nitrated, NCI initiated with discharge with thermospray introduction was tried as well as attempts to increase the sensitivity in buffer assisted ion evaporation. Due to the addition of a wire-repeller in the ion source, the sensitivities of these compounds in buffer assisted ion evaporation were comparable to those in NCI initiated by discharge. The increased mobility of the ions caused by the electric field generated by the wire-repeller appears to be the predominant mechanism in the enhancement in this type of ionization. Fink and Freas(11) found a mass dependence for the change in ion current with electric field and attributed this to ion diffusion or mobility. However, more frag-mentation was generated in NCI, and consequently the specificity was better. In order to increase the specificity for buffer assisted ion evaporation, two techniques were employed: daughter ion MS/MS experiments and repeller-induced fragmentation experi-ments. Both methods sacrificed sensitivity to some degree, but both generated useful structural information. Therefore, one would generally want to use the negative ion capabilities of thermospray to analyze for these herbicides, but in specific cases (e.g., 2,4-DB, LOD of 3.4 ng in the positive mode and LOD of 114 ng in the negative mode) the analyst would want to use the positive ion mode.

Notice

Although the research described in this article has been supported by the U.S. Environ-mental Protection Agency, it has not been subjected to Agency review and therefore does not necessarily reflect the views of the Agency. No official endorsement should be inferred.

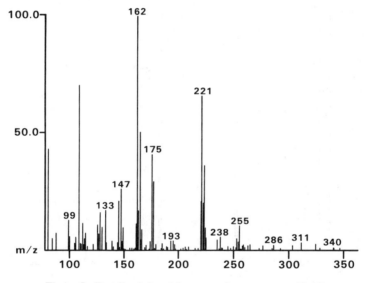

Figure 5. Repeller-induced fragmentation spectrum of 2,4-D

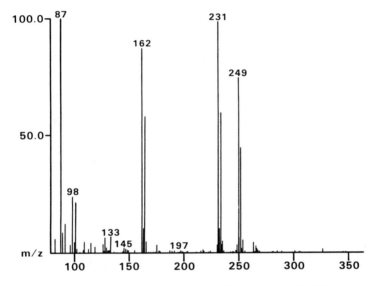

Figure 6. Repeller-induced fragmentation spectrum of 2,4-DB

TABLE IV. Positive Ion Repeller-induced Fragmentation Spectra
of Chlorinated Acid Herbicides

Compound	MW	Mass Spectra (% Relative Abundance)
Dicamba	220	$(M+H)^+(69)$, $M^+(19)$, $(MH-H_2O)^+(100)$, $(MH-H_2O-CH_3)^+(34)$, $(MH-OCH_3-OH)^+$ (27), $(C_6H_2Cl_2O)^+(29)$, $(MH-H_2O-CH_3-Cl)^+(21)$, $(C_6H_2Cl)^+(22)$
2,4-D	220	$(MH+CH_3OH)^+(5)$, $(M+NH_4)^+(7)$, $(M+H)^+(67)$, $(M-COOH)^+(42)$, $(MH-H_2O-Cl)^+(17)$, $(MH-CH_2COOH)^+(100)$, $(M-OCH_2COOH)^+(22)$, $109(70)$
MCPA	200	$(MH+CH_3OH)^+(22)$, $(M+H)^+(100)$, $(MH-H_2O)^+(16)$, $(MH-HCOOH)^+(99)$, $(MH-CH_2COOH)^+(17)$, $(C_6H_3CH_3ClO)^+(51)$, $(C_6H_3CH_3Cl)^+(51)$, $89(90)$, $90(23)$
Dichlorprop	234	$(MH+CH_3OH)^+(12)$, $(M+H)^+(56)$, $M^+(19)$, $(MH-HCOOH)^+(100)$, $(MH-CH_3CHCOOH)^+(84)$, $109(20)$
MCPP	214	$(MH+CH_3OH)^+(9)$, $(M+H)^+(48)$, $(MH-HCOOH)^+(100)$, $(MH-CH_3CHCOOH)^+(38)$, $(C_6H_3CH_3ClO)^+(21)$, $(C_6H_3CH_3Cl)^+(33)$, $(OCH(CH_3)COOH)^+(30)$
2,4,5-T	254	$(MH+CH_3OH)^+(10)$, $(M+H)^+$ $(50)^+$, $M^+(26)$, $(MH-HCOOH)^+(44)$, $(MH-CH_2COOH)^+(100)$, $(C_6H_2Cl_3)^+(15)$, $(MH-CH_2COOH-COH)^+(15)$, $145(21)$, $146(45)$, $133(35)$, $109(44)$
2,4,5-TP	268	$(MH+CH_3OH)(6)$, $(M+H)^+(30)$, $M^+(17)$, $(MH-HCOOH)^+(60)$, $(MH-CH_3CHCOOH)^+(100)$, $(MH-HCl-COOH)^+(15)$, $159(23)$, $162(19)$
2,4-DB	248	$(M+H)^+(75)$, $(MH-H_2O)^+(100)$, $(MH-(CH_2)_3COOH)^+(88)$, $101(22)$, $((CH_2)_3COOH)^+(100)$

Literature Cited

1.   Cremlyn, R. Pesticides: Preparation and Mode of Action; John Wiley and Sons:
      Chichester, 1978; p. 142.

2.   Cotterill, E. G.; Byast, T. H. "HPLC of Pesticide Residues in Environmental
      Samples." In Liquid Chromatography in Environmental Analysis; Laurence, J. F.,
      Ed.; Humana Press: Clifton, NJ, 1984.

3.   Voyksner, R. D. "Thermospray HPLC/MS for Monitoring the Environment". In
      Applications of New Mass Spectrometry Techniques in Pesticide Chemistry;
      Rosen, J. D., Ed.; John Wiley and Sons: New York, 1987.

4.   Yinon, J.; Jones, T. L.; Betowski, L. D. Rap. Comm. Mass Spectrom. 1989, 3,
      38.

5.   Betowski, L. D.; Pyle, S. M.; Ballard, J. M.; Shaul, G. M. Biomed. Environ.
      Mass Spectrom. 1987, 14, 343.

6.   Voyksner, R. D.; Bursey, J. T.; Pellizzari, E. D. J. Chromatogr. 1984, 312, 221.

7.   Geerdink, R. B.; Maris, F. A.; DeJong, G. J.; Frei, R. W.; Brinkman, U. A. Th.
      J. Chromatogr. 1987, 394, 51.

8.   Dougherty, R. C.; Roberts, J. D.; Biros, F. J. Anal. Chem. 1975, 47, 54.

9.   Yinon, J. J. Forensic Sci. 1980, 25, 401.

10.   Barceló, D. Org. Mass Spectrom 1989, in press.

11.   Fink, S. W.; Freas, R. B. Anal. Chem. 1989, 61, 2050.

12.   Bencsath, F. A.; Field, F. Anal Chem. 1988, 60, 1323.

13.   McFadden, W. H.; Lammert, S. A. J. Chromatogr. 1987, 385, 201.

14.   Safe, S.; Hutzinger, O. Mass Spectrometry of Pesticides and Pollutants; CRC
      Press, Inc: Boca Raton, FL, 1980; p. 96.

RECEIVED October 27, 1989

Chapter 6

Multiresidue Analysis of Thermally Labile Sulfonylurea Herbicides in Crops by Liquid Chromatography/Mass Spectrometry

Lamaat M. Shalaby and Stephen W. George

Agricultural Products Department, E. I. du Pont de Nemours and Company, Experimental Station, P.O. Box 80402, Wilmington, DE 19880–0402

Increasing demand for screening our environment for pesticide residues has generated a need for more efficient residue methods. Multiresidue methods, or those which measure several compounds at once, are generally more efficient than single residue methods in satisfying these needs. A mass spectrometer can be used as a universal selective detector for multiresidue analysis in different sample matrices, since it is generally blind to interferences present in the sample. In addition to selectivity, use of a mass spectrometer offers structure confirmation, and in many cases, can eliminate sample cleanup steps.

Thermospray LC/MS has been extensively used for the study of sulfonylurea herbicides (1-2). These compounds are thermally labile and can not be successfully analyzed by conventional GC/MS. Early applications of thermospray LC/MS included metabolite identification and product chemistry studies. We have recently evaluated the use of thermospray LC/MS for multi-sulfonylurea residue analysis in crops and have found the technique to meet the criteria for multiresidue methods. LC/MS offers both chromatographic separation and universal mass selectivity. Our study included optimization of the thermospray ionization and LC conditions to eliminate interferences and maximize sensitivity for trace level analysis. The target detection levels were 50 ppb in crops. Selectivity of the LC/MS technique simplified sample extraction and minimized sample clean up, which saved time and optimized recovery. Average recovery for these compounds in crop was above 85%.

This report describes the application of LC/MS for residue analysis of three sulfonylureas in wheat grain. An acetonitrile/water mixture was used for extraction and followed by centrifugation and filtration. No further clean up of the wheat grain extract was needed prior to chromatographic and spectroscopic analysis.

EXPERIMENTAL

The HPLC system consisted of a Varian Model 5560 liquid chromatograph equipped with a constant-flow pump, a variable wavelength detector (Varian, Instrument Group/Walnut Creek, CA), a Rheodyne injector valve and an Alltech Spherisorb@ ODS column, 4.6 mm id x 25 cm (Alltech/Applied Science, IL). The mass spectrometer was a Finnigan Model 4600 quadrupole instrument with

0097–6156/90/0420–0075$06.00/0

the INCOS Data System (Finnigan MAT, CA). The LC/MS interface was a Vestec thermospray with a discharge electrode and filament ionization (Vestec Corporation, TX). The 0.5M ammonium acetate solution was added postcolumn by the dual-piston pulseless Kratos Model Spectroflow 400 HPLC pump (ABI Analytical Kratos Division). A pulseless HPLC pump is essential with thermospray LC/MS to maintain a stable ion signal. The LC/MS system is equipped with two 2 μm on-line filters (Kel-F ring-A-101X, Thomson Instrument Co., DE 19711) to prevent clogging of the capillary LC/MS interface line.

The wheat grain samples were extracted using a Vortex mixer, (Thermolyne Corporation, Iowa) and an ultrasonic bath, (Branson Cleaning Equipment Co., CT). The wheat grain extracts were centrifuged using International Clinical Centrifuge (International Equipment Co., Mass). Aliquots of the filtered (Gelman Acrodisc-CR Filters, 0.45 micron, (Gelman Sciences, MI), extracts were evaporated by an N-Evap Analytical Evaporator, (Organomation Association, MA).

The solvents were HPLC grade acetonitrile, EM OMNISOLV solvent, (EM Science, N.J.) and distilled, deionized water using a MILLI-Q water Purification System (Millipore Corp., MA). The ammonium acetate used to prepare the 0.5M solution which is added postcolumn to aid the thermospray ionization was 'Baker Analyzed' Reagent (J. T. Baker, NJ).

The herbicides used in this study were standard reference materials (Du Pont Company, Wilmington, DE)

A 1 μg/mL stock standard solution of each herbicide was prepared in acetonitrile. This solution was kept refrigerated and used for both sample fortification and preparation of the LC/MS calibration solutions. The calibration solutions were prepared by diluting an aliquot of the stock solution with a solution of 40/60 acetonitrile/water, v/v.

RESULTS AND DISCUSSION

General LC/MS Conditions for Sulfonylurea Multiresidue Analysis. We have developed general thermospray LC/MS conditions for the purpose of separating and detecting six different sulfonylurea herbicides. These conditions can be used as a guide for a variety of LC/MS residue applications which may require the analysis of one or more of these herbicides. Our procedure includes GLEAN (chlorsulfuron), ALLY (metsulfuron methyl), HARMONY (thiameturon) and EXPRESS cereal herbicides, CLASSIC (chlorimuron ethyl) soybean herbicide and OUST (sulfometuron-methyl) noncrop land herbicide. (Structure 1)

Figure 1 shows the LC/MS thermospray total ion chromatogram of 0.25 μg standard mixture of the six sulfonylurea herbicides. Gradient HPLC conditions were used to separate the six compounds in less than 25 minutes total run time. The mobile phase composition was kept isocratic at 30% acetonitrile/.05M formic acid for the first 15 minutes to separate the four herbicides HARMONY, ALLY, OUST and GLEAN. A gradient from 30% acetonitrile to 60% in 10 minutes was then used to elute EXPRESS and CLASSIC. An acidified mobile phase is used with sulfonylureas to keep them in the undissociated form which is retained on the HPLC column (3). Organic acids are recommended for use with LC/MS to prevent the formation of deposits in the mass spectrometer source and to prevent clogging of the thermospray interface probe tip. In this work we used formic acid.

Ammonium acetate, needed for thermospray ionization, is added post-column so as not to affect sulfonylurea retention on the LC column. The post-

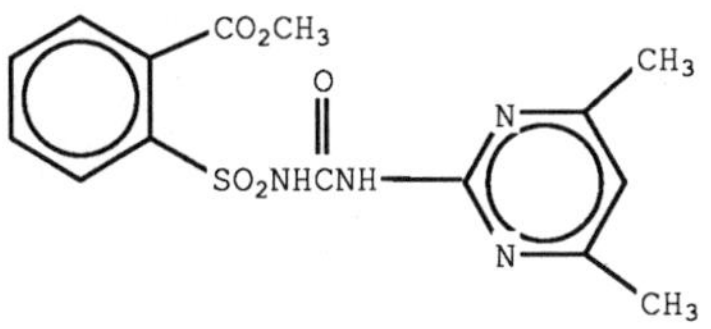

OUST

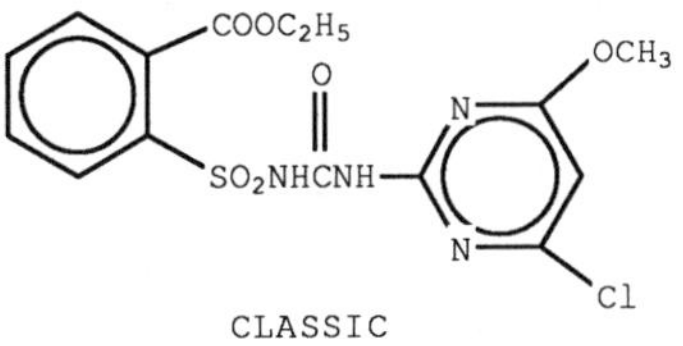

CLASSIC

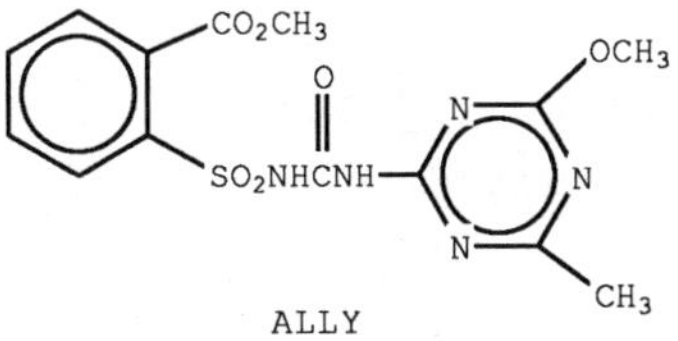

ALLY

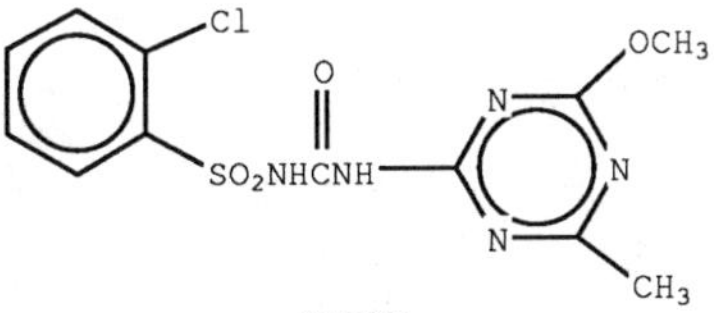

GLEAN

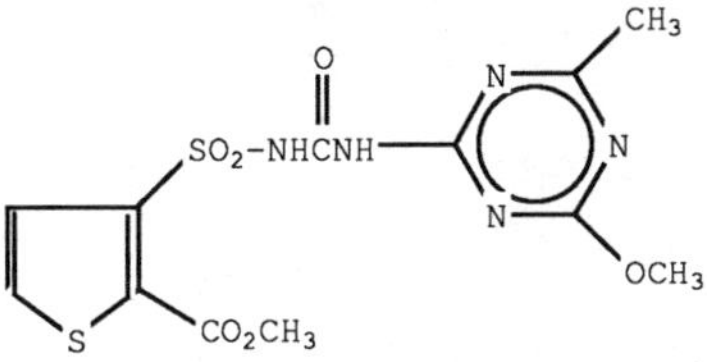

HARMONY

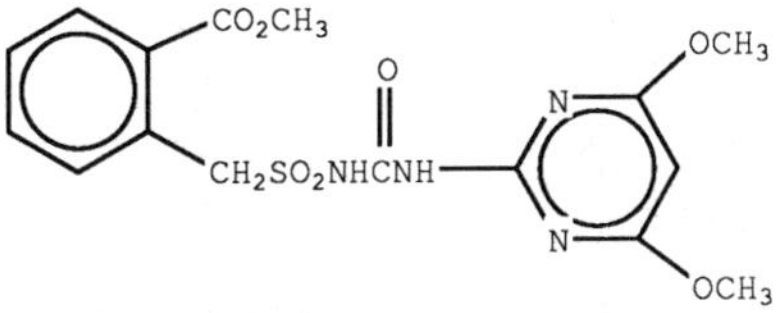

LONDAX

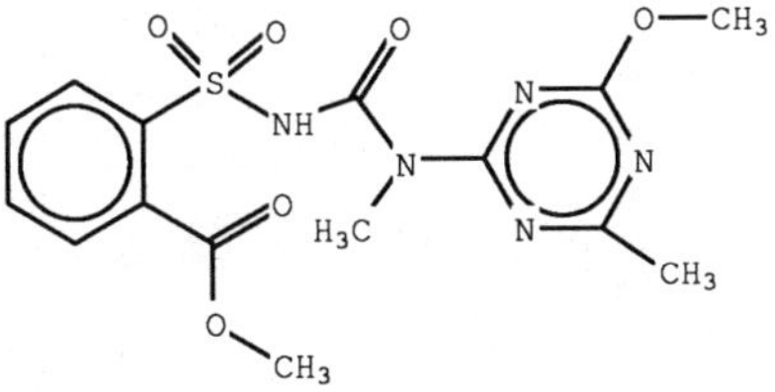

EXPRESS

Structure 1

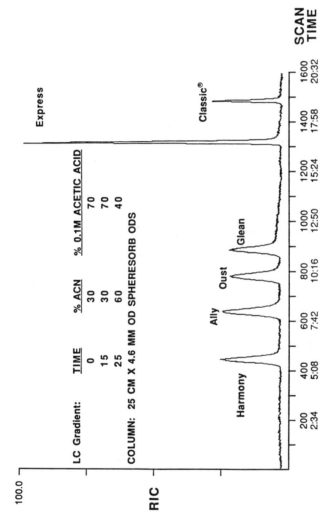

Figure 1. Thermospray LC/MS total ion chromatogram of 0.25 μg standard mixture of six sulfonylurea herbicides.

column added solution will be diluted on-line as it mixes in a stainless steel Valco low-dead volume tee with the mobile phase to give a final concentration of 0.1 M ammonium acetate.

High concentration of organic modifier in the thermospray solvent has a quenching effect on the ion signal and is more apparent with acetonitrile than with methanol (4-5). With gradient LC, the ion signal generally decreases as the organic modifier increases. Decreasing the vaporizer tip temperature during the gradient will minimize the loss in ion signal but it will not eliminate it (6). We normally use a narrow-range LC gradient (a total of 30-40% increase in the organic modifier co: .ntration) to minimizes this effect. In this multiresidue separation, the thermospray ion signal was compensated for the gradient LC by programing the vaporizer temperature down as the acetonitrile concentration increases (7). (Structure 1)

A typical thermospray ionization mass spectrum for a sulfonylurea contains a weak protonated molecular ion and three to four characteristic fragment ions. Figure 2 shows the thermospray positive ion mass spectrum for HARMONY. The spectrum contains a protonated molecular ion at m/z 388, the sulfonamide ammonium adduct ion at m/z 239 and the protonated triazine urea fragment ion at m/z 184. At the same time, Figure 3 shows the positive ion thermospray mass spectrum for LONDAX. It contains the protonated pyrimidine amine at m/z 156, the protonated pyrimidine urea is at m/z 199 and the sulfonamide ammonium adduct ion at m/z 247. LONDAX (bensulfuron methyl) is a sulfonylurea rice herbicide and it elutes between EXPRESS and CLASSIC if we use the LC conditions outlined in Figure 1. HARMONY and LONDAX thermospray spectra were generated with the thermospray vaporizer tip temperature at 150°C and the source block temperature at 320°C.

For the six sulfonylurea herbicides included in Figure 1, we monitored the protonated molecular ion for each herbicide in addition to one or two major fragment ions. Table 1 shows the ions selected for each of the sulfonylurea and Figure 4 shows the ion traces for each compound. The table shows two ions which are common for some of these sulfonylurea herbicides. HARMONY, ALLY and GLEAN contain the same triazine urea ion at m/z 184 while ALLY, OUST and EXPRESS contain the same sulfonamide ions at m/z 233. Selecting these common ions for quantitation will increase the overall sensitivity for multiresidue analysis.

Table 1
Selected Ions for Six Sulfonylureas

Herbicide				Selected Ions(m/z)					
	155	165	184	203	233	358	382	388	415
HARMONY			+					+[MH]	
ALLY			+		+		+[MH]		
OUST		+			+				
GLEAN			+			+[MH]			
EXPRESS	+				+				
CLASSIC				+					+[MH]

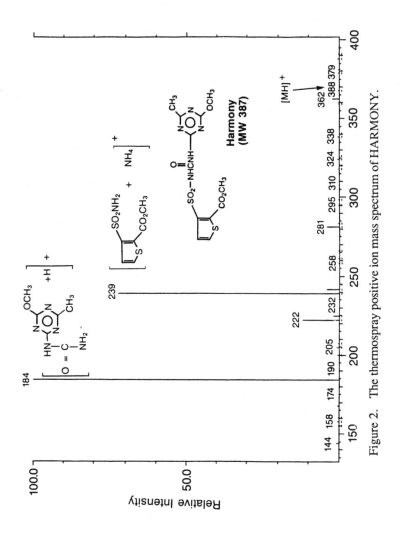

Figure 2.   The thermospray positive ion mass spectrum of HARMONY.

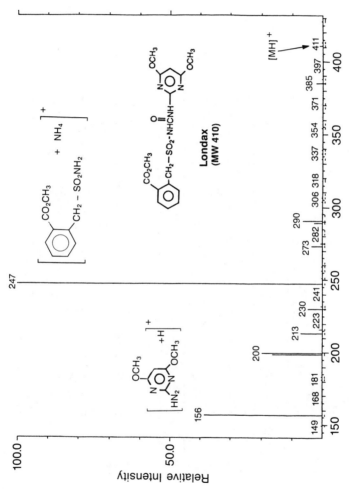

Figure 3.   The thermospray positive ion mass spectrum of LONDAX.

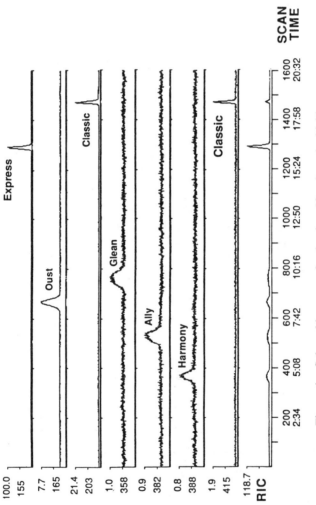

Figure 4. Selected ion traces for the six sulfonylurea herbicides.

The LC/MS separation and detection shown in this method illustrate the multi-residue capabilities of LC/MS. The combination of the LC separation for six sulfonylurea herbicides within 25 minutes and the selectivity and structure confirmation of the mass selective detector offer the basis for efficient residue analysis. LC/MS residue methodology can be a cost effective alternative to conventional residue methodology. The mass selectivity will simplify the extraction procedure and minimize sample clean up. A simplified sample preparation saves time and reduces analyte loss during sample preparation which will maximizes recovery.

Application of LC/MS to Sulfonylurea Multiresidue Analysis in Wheat Grain. To test the applicability of this LC/MS method for residue analysis, we applied the method for the analysis of HARMONY, GLEAN and LONDAX in wheat grain. The objective was to determine the effect of the matrix on the separation and ion signal. Another objective was to determine the recovery using a simple extraction procedure with no sample clean up.

The samples were prepared by weighing 10 g of ground grain and fortifying with the 1μg/mL standard mixture of HARMONY, GLEAN and LONDAX. Fortification was carried out in duplicate at 0.05, 0.20 and 0.50 ppm, and solvents were evaporated under a stream of nitrogen. Each sample was extracted twice, once with 20 mL and again with 10 mL of 80% acetonitrile/water extraction solvent. Samples were allowed to stand for several minutes, vortexed, ultrasonicated for 10 minutes, vortexed and centrifuged at 2000 rpm for 15 minutes. The supernatant from each sample was decanted and the total volume recovered for each sample was recorded (28-30 mL). An aliquot of 10 mL of the filtered extract was reduced to 0.5 mL and adjusted to 1 mL by the addition of acetonitrile.

The purpose of the last step is to keep the final sample solvent of similar composition to the LC mobile phase to maintain the sulfonylurea LC peak shape. Standards were prepared at the 0.1 - 2.0 μg/mL range in 50% acetonitrile/water.

Thermospray Operation. Figures 2 and 3 show the thermospray positive ion mass spectra of HARMONY and LONDAX. The spectra contain three to four structurally informative ions. For quantitation at the method detection level in each case, we selected the most intense ions.

The back pressure (35-45 bar) generated from the thermospray evaporation process in the capillary interface line is monitored by the Spectroflow pump used for the post column addition of the 0.5 M ammonium acetate solution. It should be stable($\pm$ 1 bar) to insure good reproducibility of the ion signal. An increase in the back pressure would indicate partial clogging of the in-line filter or the thermospray probe tip. The blockage must be eliminated before proceeding with the analysis.

In order to minimize the mass spectrometer source contamination from residue analysis and to prevent clogging the LC/MS vaporizer tip we vent the LC effluent for the first 5 min. The effluent is switched back to the mass spectro-meter at least 5 min before the first peak elutes to allow equilibration of the ionization process and maintain reproducibility.

The thermospray vaporizer control temperature (T1) is the critical parameter for total ion sensitivity. A temperature study was performed initially using a standard solution, injected by flow injection, without the column. Using a temperature ramp and monitoring the ion signal with multiple standard injections, we determined the optimum temperature for the specific compound. Figure 5

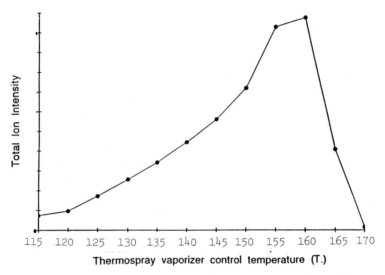

Figure 5.   Effect of thermospray vaporizer tip temperature on the total ion signal for a sulfonylurea herbicide.

shows the effect of thermospray vaporizer temperature on a typical sulfonylurea total ion signal. The optimum temperature range is 5-7°C wide, keeping the probe tip temperature at few degrees lower than the maximum minimizes the accumulation of deposits in the probe tip from the grain extract. This can cause high back pressure which will reduce sensitivity and reproducibility.

Instrumental Set Up

LC Column:	Spherisorb ODS 25 cm x 4.6 mm od
Flow rate:	1.2 mL/min on column
Post Column	
Addition:	0.3 mL/min of 0.5 M ammonium acetate
Mobile Phase:	Gradient

Time (min)	%ACN	%0.1M acetic acid
0	30	70
5	30	70
15	45	55

Injection Size:	50 μL
Retention Time:	HARMONY at 15.5 min
	GLEAN at 16.5 min
	LONDAX at 22.2 min
Selected ion	
monitored:	m/z 184 and 247
Ionization Mode:	thermospray
Mass	
Calibration:	polypropylene glycol (PPG) with the thermospray LC/MS source.
Calibration	0.1, 0.2, 0.5, 1.0 and 2.0 μg/mL standard
Solutions:	mixture of HARMONY, GLEAN and LONDAX in ACN/H$_2$O
Electron Multiplier	
voltage:	1000 V

CALCULATIONS

The amount of each herbicide recovered from the treated wheat grain as detected by LC/MS is calculated as shown:

$$\text{ppm (}\mu\text{g herbicide/g wheat grain) found} = \frac{(a)\ (tv)}{(rf)\ (cf)\ (w)}$$

where

(a)   is the peak area (ion counts)

(tv) is the total volume of sample extract

(rf) is the response factor derived from the peak area of a standard divided by the standard concentration (ion counts/μg/mL)

(cf)   is the concentration factor derived from the volume
       of the extract before concentration divided by the volume of the extract after
       concentration

(w)   is the original weight of the wheat grain sample (g)

$$\% \text{ recovery} = \frac{\text{ppm found}}{\text{ppm added}} \times 100\%$$

Figure 6 shows the thermospray LC/MS chromatogram for a standard mixture
of 25 ng HARMONY, GLEAN, and LONDAX. The chromatogram indicates
good separation, stable ion signal and good sensitivity. Figure 7 shows a
representative linear calibration curves for GLEAN, HARMONY and LONDAX
in the range which corresponds to the amount of herbicide injected to the LC/MS
with the 0.05-0.5 ppm fortified samples.
      Figure 8 shows the LC/MS chromatograms of an untreated wheat grain
analyzed for the three sulfonylurea herbicides. The chromatogram indicates no
interferences. Figure 9 shows the LC/MS chromatograms of fortified wheat
grain at 0.1 ppm with the three herbicides. The three herbicide peaks were
detected without interferences especially in the selected ion trace as shown.
      The recovery study results for the three sulfonylurea herbicides in wheat
grain are shown in Table 2. Untreated wheat grain controls were fortified with
HARMONY, GLEAN and LONDAX at 0.05, 0.20 and 0.50 ppm. The average
recovery for HARMONY was 90%, for GLEAN was 75%, and for LONDAX
was 95%.

Table 2
Recovery Results for Multi-Sulfonylurea Residues in Wheat Germ

Herbicide	Fortification	Average % recovery
HARMONY	50 - 500 ppb	90
GLEAN	50 - 500 ppb	75
LONDAX	50 - 500 ppb	95

      This method relies on the mass selectivity of the mass spectrometer in
addition to the LC separation. No interferences were detected with any of these
selected ions. The ions selected for quantitation were the most intense ions to
obtain good sensitivity. Structure confirmation is inherent in an LC/MS method
due to the structural information of the selected ion and their relative abundance.
In addition LC/MS offers the LC chromatographic confirmation based on
retention time relative to the reference standard.
      In this method we eliminated sample clean up and used a simple extraction
procedure. Sample throughput is five times higher than conventional methods
where multiple clean up steps are needed to eliminate interferences. LC/MS
selectivity led to a much faster method development (few weeks) compared to the
lengthy conventional methods (few months). A conventional method with
conventional nonselective detector such as the UV relies on extracting the matrix
interference before analysis. This requires evaluating different extraction

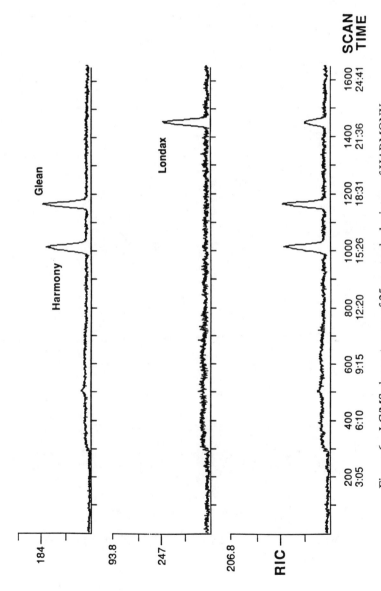

Figure 6. LC/MS chromatogram of 25 ng standard mixture of HARMONY, GLEAN and LONDAX.

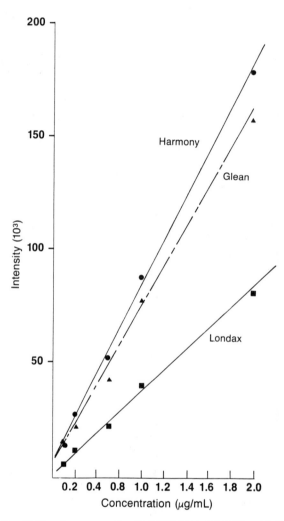

Figure 7.   Calibration curves for GLEAN, HARMONY and LONDAX.

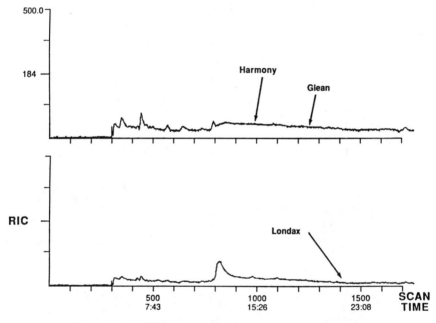

Figure 8.   LC/MS Chromatogram of untreated wheat grain.

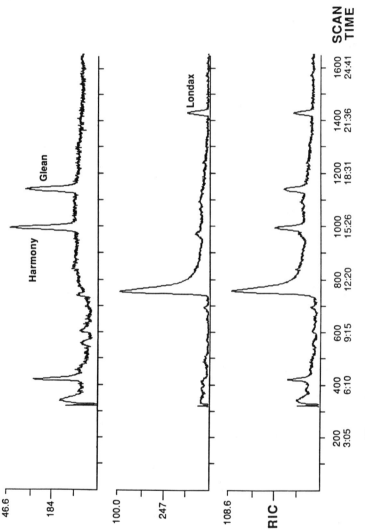

Figure 9. LC/MS Chromatogram of fortified wheat grain at 0.1 ppm
HARMONY, GLEAN and LONDAX.

solvents and solid phase extraction which is time consuming and may lead to a low recovery of the analyte. With the mass selective detector combined on line with LC, only the targeted compound will be detected.

CONCLUSION

LC/MS has emerged as a sensitive and selective residue methodology for the trace organic analysis of crop protection chemicals. This technology is especially applicable to low application rate herbicides such as sulfonylureas because it requires minimal sample processing and clean-up prior to chromatographic and spectroscopic quantitation.

The thermospray LC/MS with selected ion monitoring is applicable to multiresidue sulfonylurea herbicides. This method offers simultaneous extraction and analysis of three herbicide compounds in a fast and efficient way, in addition to good recovery.

REFERENCES

1.   L. M. Shalaby, Chapter 12, Applications of New Mass Spectrometry Techniques in Pesticide Chemistry (Ed. J. B. Rosen), 91 in Chemical Analysis, Wiley-Interscience, New York, p. 161 (1987).

2.   L. M. Shalaby and R. W. Reiser, Chapter 11, Mass Spectrometry of Biological Materials, (Ed C. McEwen and B. Larsen), Marcel Dekker, in print.

3.   E. M. Beyer, M. J. Duffy, J. V. Hay, D. D. Schlueter, chapter 3, Herbicides : Chemistry, Degradation, and Mode of Actions Vol. 3, Marcel Dekker, Inc. Chapter 3 (1987).

4.   L. M. Shalaby, Applications of Gradient Elution LC/MS for the Analysis of Agricultural Compounds, 33rd Annual Conference of Mass Spectrometry and Allied Topics, San Diego, CA, 1985.

5.   L. M. Shalaby, Trace Level Analysis of Thermally Labile Herbicides by on-Line Thermospray LC/MS, 33rd Annual Conference on Mass spectrometry and Allied Topics, San Diego, CA, 1985

6.   Garteiz and M. L. Vestal, LC Magazine, 3(4), 334 (1985).

7.   Vestec Thermospray LC/MS Interface, Instruction/maintenance manual, Finnigan Edition, Vestec Corporation, Houston, TX.

RECEIVED November 9, 1989

Chapter 7

Applications of Liquid Chromatography/Negative Ion Mass Spectrometry in Studies of Herbicide Metabolism

R. T. Solsten, H. Fujiwara, and E. W. Logusch

Life Sciences Research Center, Monsanto Agricultural Company, Chesterfield, MO 63198

The combination of electron capture ionization with liquid chromatography/mass spectrometry produces a valuable technique for metabolism studies of fluorinated compounds. Selective ionization of halogenated components simplifies sample purification since most biological constituents are transparent to detection. This report details the application of this technique to in vitro and in vivo metabolism studies of fluorinated herbicides.

The technique of electron capture gas chromatography/mass spectrometry (GC/MS) is widely used for the analysis of halogenated compounds because it frequently provides a substantial increase in sensitivity over positive ion chemical ionization mass spectrometry (1-3). The combination of electron capture ionization with liquid chromatography/mass spectrometry (LC/MS) offers several advantages over conventional GC/MS and positive ion LC/MS. Since LC/MS is a relatively soft ionization technique, mass spectra can be obtained directly from a biological matrix without the need for isolation or derivatization of the analyte. Selective ionization of electronegative compounds renders most biological constituents transparent to detection, resulting in less complicated chromatograms and allowing rapid identification of peaks associated with the analyte. A 100- to 1000-fold increase in sensitivity is often observed for compounds capable of electron capture (4,5). This enhanced sensitivity permits the routine detection and identification of nanogram quantities of halogenated components from crude biological matrices without purification. In addition, electron

0097–6156/90/0420–0092$08.75/0
© 1990 American Chemical Society

capture ionization produces odd-electron molecular ions which can undergo fragmentation pathways not available to positive ions. This increased fragmentation frequently provides additional structural information which is complementary to that obtained from positive ion chemical ionization (6).

We have found that several new fluorinated herbicide candidates currently under investigation at Monsanto are very efficient at electron capture and thus are ideally suited for negative ion LC/MS. One such candidate is the pre-emergence herbicide dithiopyr 1 which is illustrated in Figure 1. Dithiopyr is representative of a novel class of fluorinated pyridine herbicides whose metabolism has been studied extensively by negative ion LC/MS. This report describes several applications of negative ion LC/MS to in vitro and in vivo metabolism studies of these fluorinated herbicides.

Experimental

Samples and solvents. The experimental herbicides and their metabolites were prepared in the research laboratories of Monsanto Agricultural Company. Mobile phase solvents consisted of HPLC-grade acetonitrile or methanol (Fisher Scientific, Fair Lawn, New Jersey), and a 1% solution of formic acid (EM Science, Cherry Hill, New Jersey) in purified "Milli-Q" water (Waters Associates, Milford, Massachusetts).

Preparation of in vitro metabolite samples. Incubations were performed by combining the following components: 0.5 mL of the S9 fraction of rat liver microsomes; 1 mg of ß-NADPH; 2 μg of glucose-6-phosphate, 5 μL of glucose-6-phosphate dehydrogenase; 1.5 mg of glutathione (Sigma Chemical Co., St. Louis, Missouri); 5 μg MgCl$_2$; 0.5 mL of 0.4 M phosphate buffer, pH=7.4; 3 μg of substrate. The reaction mixture was stirred in a water bath maintained at 37 degrees C. and aliquots were removed frequently to monitor the progress of the reaction. At each sampling time an aliquot was added to an equal volume of methanol in a microcentrifuge tube to quench the reaction. The tube was then centrifuged to remove insoluble proteins. The progress of the incubation was monitored by HPLC and on-line radioactivity flow detection.

Preparation of urine samples. Urine was collected from Fischer 344 rats eight hours after dosing with the herbicide candidate. The urine was kept frozen until mass spectral analysis. After thawing, the urine was centrifuged for two minutes at 13,600 g before injection onto the liquid chromatograph.

Preparation of fecal extracts. Feces were collected from
Fischer 344 rats for one hundred and twenty hours after
dosing with the herbicide candidate. A sample of one gram
of the pooled fecal material was homogenized with 9 mL of
water. The homogenate was diluted with 10 mL of acetonitrile and the mixture was shaken for 20 minutes. The
supernatant was decanted and the feces were extracted
again with acetonitrile/water (70/30, v/v). After centrifugation, the combined extracts were partitioned with an
equal volume of methylene chloride. The aqueous fraction
was centrifuged at 13,600 g for two minutes before injection onto the liquid chromatograph.

Liquid chromatography/mass spectrometry. Liquid chromatography coupled with mass spectrometry was performed
using a system assembled from components manufactured by
Waters Associates (two Model 510 pumps and a Model 680
gradient controller). The samples were introduced into
the mass spectrometer after elution from Waters μBondapak
C18 (3.9 X 300 mm) or Beckman Ultrashpere ODS (4.6 X 250
mm) columns. The mobile phase consisted of a linear
gradient of: a) 5% to 95% acetonitrile in 1% aqueous
formic acid over 20 min at a flow rate of 1.0 mL/min; or
b) 5% to 95% methanol in 1% aqueous formic acid over 30
min at a flow rate of 0.8 mL/min. HPLC columns were
protected by precolumns (Brownlee Labs RP-18 Newguard
7μ, 3.2 X 15 mm). The liquid chromatograph effluent was
introduced into a Finnigan 4535 quadrupole mass spectrometer by means of a Vestec Model 701 thermospray source
operated in the discharge ionization mode. The source
block was maintained at 275 to 280 °C and the tip heater
was maintained at 310 to 320 °C throughout the analyses.
The vaporizer temperature was controlled by a gradient
compensator and typically decreased from 160 to 120 °C
during the gradient progression. Both positive and negative ions were analyzed. The mass spectrometer was
scanned from 200 to 700 amu in 1.5 sec and the data were
processed with a Data General Nova 4 computer using INCOS
software.

Fast atom bombardment mass spectrometry. Fast atom bom-
bardment/mass spectrometry (FAB/MS) analyses were performed on a VG ZAB-HF mass spectrometer equipped with an
Ion Tech fast atom gun. Xenon gas was activated to 8 kV
and 1.5 mA ion current for the fast atom generation. An
accelerating voltage of 8 kV was applied to the FAB
source. The mass spectrometer was scanned from 800 to
80 amu using an exponential down scan mode at 5 seconds
per decade with a 1 second interscan time. The data were
recorded with a PDP 11/24 computer and were processed with
VG 11/250 software.

Results and Discussion

Sensitivity of Electron Capture Ionization. We have found that negative ion LC/MS provides substantial increases in sensitivity as compared with positive ion LC/MS during analysis of fluorinated pyridines such as dithiopyr 1. An example of such sensitivity enhancement is shown in Figure 2. The top panel illustrates the total ion current trace obtained from injection of increasing amounts of dithiopyr in the positive ion mode with discharge ionization. The detection limit for 1 is approximately 1000 nanograms for positive ion LC/MS analysis. The bottom trace shows the total ion current observed for the same sample size under negative ion conditions. The detection limit for dithiopyr is significantly below the 100 nanogram level shown in Figure 2 and would likely be in the low picogram range if multiple ion detection were employed.

Increased Fragmentation During Negative Ion Analysis. The formation of odd electron ions during electron capture generally results in increased fragmentation for the fluorinated pyridines when compared with positive ion chemical ionization. This characteristic frequently provides additional structural information and expedites metabolite identification. The extent of fragmentation typically observed for this class of compounds is illustrated in Figure 3 with the positive and negative ion mass spectra of a common metabolite of dithiopyr, the diacid 2.
 The positive ion spectrum (upper panel) displays only a protonated molecular ion with negligible fragmentation. The negative ion spectrum (lower panel) exhibits an ion for the deprotonated molecule at m/z 340 as well as several diagnostic ions arising from successive losses of HF. An overall reduction of the species represented by m/z 301 appears to occur via replacement of a fluorine atom by a hydrogen atom from the mobile phase producing the weak ion at m/z 283(7). The details of this process were established by conducting negative ion analysis of the diacid 2 in a deuterium oxide mobile phase. Under these conditions the ion at m/z 283 shifts to m/z 284, confirming the replacement of a fluorine atom by deuterium. The successive losses of HF presumably account for formation of a five-membered lactone on each side of the pyridine ring as shown in Scheme 1. The ions associated with these losses (m/z 301 and m/z 321) were observed at m/z 301 and m/z 322 in the deuterium oxide analysis, indicating the involvement of exchangeable protons in this process. Such losses of HF make it possible to distinguish monoacid and diacid metabolites of dithiopyr, since monoacids undergo only a single loss of HF prior to the reductive loss of fluorine.
 The significant differences observed in the positive and negative ion mass spectra of fluorinated pyridines greatly assist metabolite identification in biological

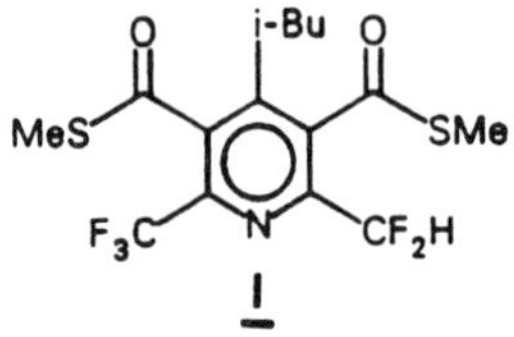

Figure 1.   Structure of dithiopyr **1**.

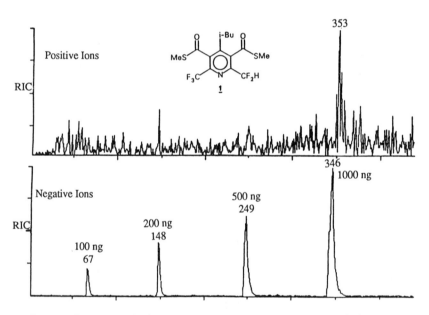

Figure 2.   Total ion current traces from positive and negative ion LC/MS analyses of dithiopyr **1**.

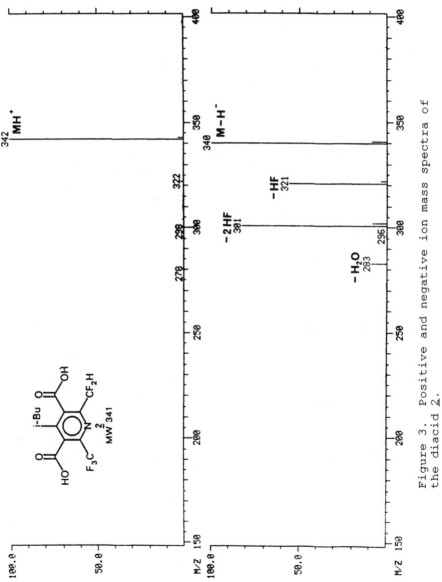

Figure 3. Positive and negative ion mass spectra of the diacid 2.

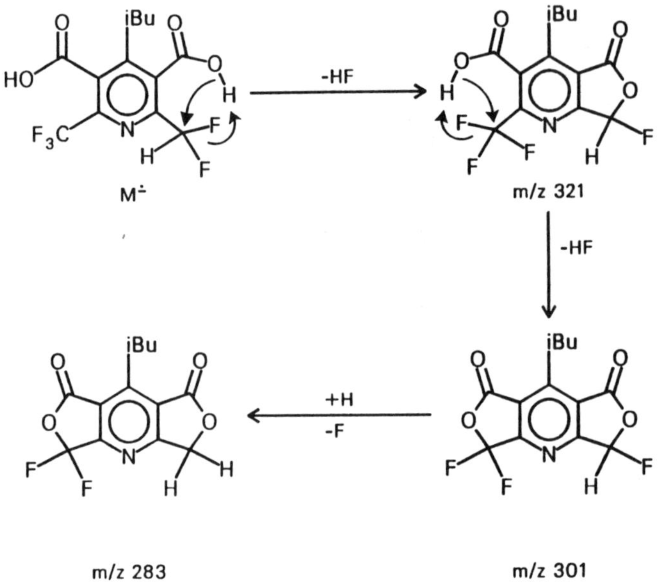

Scheme 1. Fragmentation of diacid 2 during negative ion analysis.

matrices. Analysis of crude microsomal reaction mixtures
or urine samples with the pulsed positive ion negative
ion chemical ionization (PPINICI,8) technique provides
complementary information which expedites metabolite
identification. Selective ionization of fluorinated
components by electron capture allows rapid location of
metabolite peaks in the negative ion total ion current
chromatogram. Molecular weight information for each metabolite is provided by the positive ion mass spectrum,
while diagnostic fragment ions are obtained from the
negative ion mass spectrum. The PPINICI experiment, when
coupled with LC/MS, permits a rapid survey of metabolites
present in a crude biological sample without the need for
extensive purification or derivatization.

Analysis of In Vitro Metabolism Reactions by LC/MS. An
increasingly useful approach in metabolism studies
involves a preliminary in vitro metabolism study using
liver or kidney microsomal preparations (9,10). When combined with LC/MS, such a preliminary study can rapidly
provide information about potential metabolites expected
from subsequent in vivo studies. A flow chart of a typical experimental procedure for the microsomal reaction
with mass spectral identification is shown in Figure 4.
The herbicide substrates are generally labeled with both
[13]C and ^{14}C, which allows monitoring of the reaction by
radioactivity detection, and facilitates metabolite identification based on characteristic doublet ions in the
mass spectra. The entire procedure can be completed in
several hours on a microgram scale, generating a survey of
potential metabolites.

We have found it useful to analyze such microsomal
reactions by PPINICI taking advantage of the selectivity
of electron capture ionization while obtaining confirmation of molecular weight assignments by the positive
ion mass spectrum. The total ion current traces from
PPINICI LC/MS analysis of the microsomal incubation of
dithiopyr are shown in Figure 5. The negative ion trace
(lower panel) is substantially less complicated than the
positive ion trace and allows rapid location of the metabolites. The major products observed in this reaction
were the cysteinylglycine conjugate 4 and the glutathione
conjugate 5. Other reaction products were the isomeric
monoacids 6 and 7. This analysis required only twenty
minutes to complete, and a kinetic study with half-hour
sampling points could be easily performed without the need
for quenching or freezing of samples.

The positive and negative ion mass spectra of metabolite 5 are shown in Figure 6. The starting material in
this experiment was labeled only with ^{14}C in the pyridine
ring so that any ions associated with a metabolite will
display a M+2 doublet. The highest-mass ion in the positive ion spectrum is found at m/z 532 and arises via loss
of glutamic acid from the protonated molecular ion. This

Combine S-9 fraction of rat liver microsomes
with NADPH, glucose-6-phosphate, MgCl$_2$,
glucose-6-phosphate dehydrogenase, and
glutathione

↓

Add labeled compound of interest and
incubate at 37 ° C

↓

Remove aliquot, quench with methanol, and
centrifuge for two minutes at 13,600 g

↓

Inject supernatant onto LC column and
acquire PPINICI mass spectra

Figure 4.   Flowchart for LC/MS analysis of in vitro
metabolites.

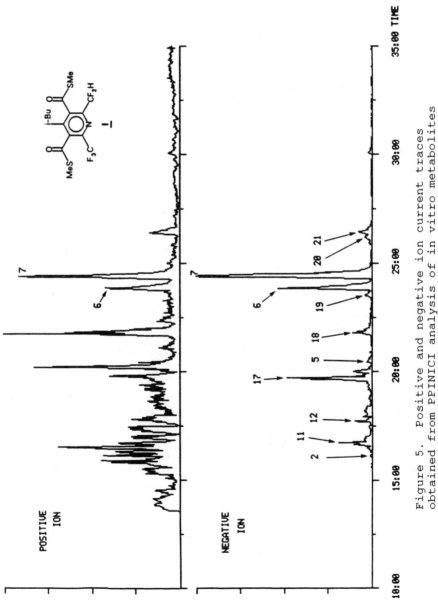

Figure 5. Positive and negative ion current traces obtained from PPINICI analysis of in vitro metabolites of dithiopyr.

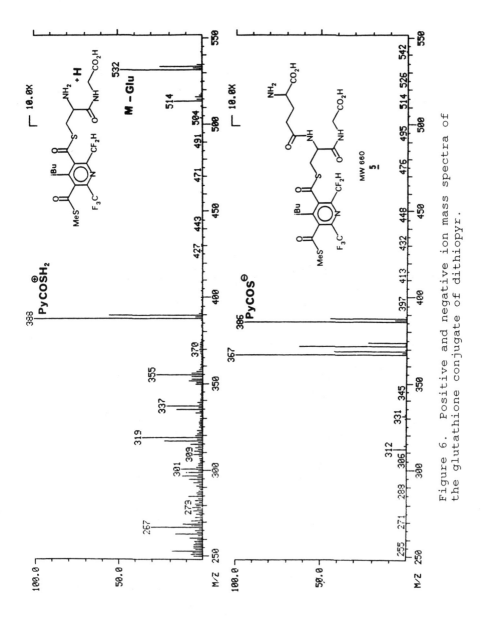

Figure 6. Positive and negative ion mass spectra of the glutathione conjugate of dithiopyr.

fragmentation pathway has previously been reported in
thermospray mass spectra of glutathione conjugates of a
series of aromatic compounds (11,12). The most abundant
ion at m/z 388 is due to a protonated thioacid which
arises from loss of the peptide moiety. The anion of this
thioacid is frequently observed in negative ion thermospray mass spectra of peptide conjugates (13) and is found
at m/z 386 in the negative ion mass spectrum of 5. From
these data we can see that the positive ion spectrum provides molecular weight information while the negative ion
spectrum provides characteristic fragmentation patterns
useful in structure elucidation. These structure assignments were confirmed by comparison with retention times of
authentic samples which had been previously characterized
by derivatization and FAB mass spectral analysis.

The positive and negative ion mass spectra of metabolite 7 are shown in Figure 7. Once again, the positive
ion spectrum simply provides the molecular weight of the
metabolite with little fragmentation. The negative ion
spectrum features a characteristic loss of HF from the
molecular anion giving rise to the base ion at m/z 351.
Since metabolite 7 is a monoacid, only one loss of HF
occurs prior to the final reductive loss of fluorine which
produces the ion at m/z 333. This fragmentation con-
trasts with that of the diacid 2, where the lactonization
of each carboxylate results in the successive losses of
HF, followed by a reductive loss of fluorine (Scheme 1).

Analysis of In Vivo Metabolites by LC/MS. Aside from its
utility in the rapid analysis of in vitro experiments,
negative ion LC/MS is an invaluable tool for in vivo metabolism studies of fluorinated herbicides. Such studies
are often preceded by a quick survey of urinary or fecal
metabolites by LC/MS. The individual components can be
isolated for further characterization by derivatization
or high resolution FAB mass spectral analysis. Urine
samples can be analyzed directly by LC/MS following centrifugation or filtration to remove suspended solids.
Figure 8 illustrates the total ion current trace corresponding to negative ion LC/MS analysis of urine obtained
from rats four hours after dosing with dithiopyr 1. As
was observed in the microsomal metabolism study, the
parent compound is initially metabolized in rats primarily
via hydrolysis to the isomeric monoacids 6 and 7. Evidence of further hydrolysis to the diacids 2,8, and 9, is
observed in the urine, but negligible glutathione or other
peptide conjugation is detected at this early timepoint.
Based on radioactivity measurement, the total amount of
metabolite sample in the analysis was 16 micrograms. The
two monoacids 6 and 7 accounted for about half of the
total radioactivity, indicating that many of the minor
metabolites were detectable at the sub-microgram level.
The mass spectra of the diacids 2 and 8 (Figure 9) display the expected successive losses of HF resulting in the

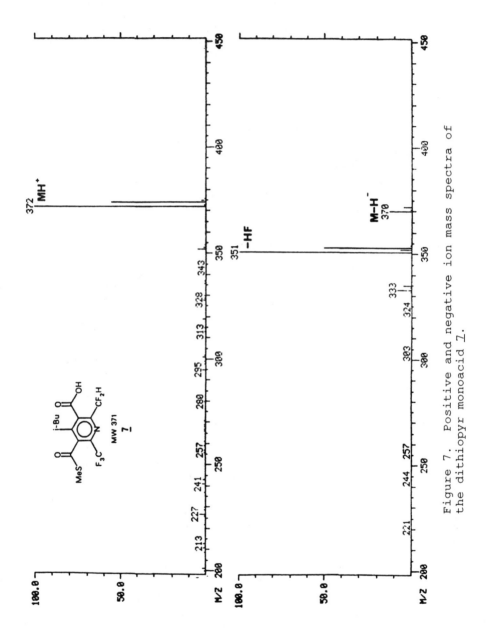

Figure 7. Positive and negative ion mass spectra of
the dithiopyr monoacid 7.

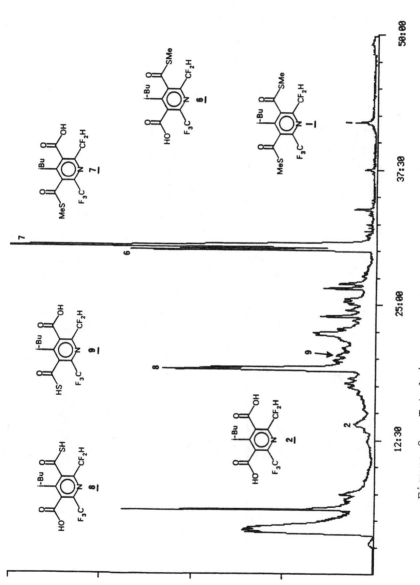

Figure 8. Total ion current trace from negative ion LC/MS analysis of dithiopyr urinary metabolites.

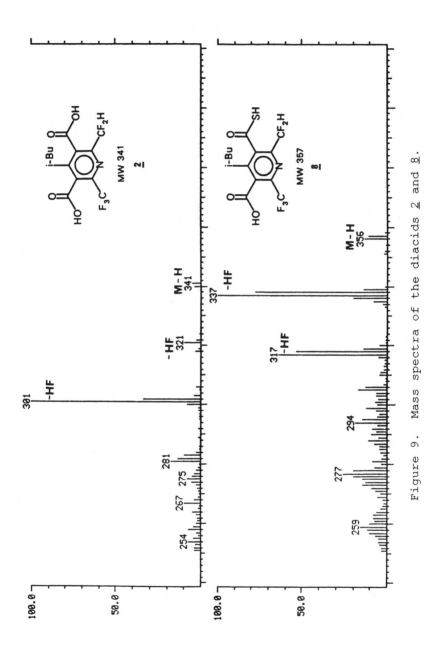

Figure 9.   Mass spectra of the diacids 2 and 8.

formation of a five-membered lactone on each side of the
pyridine ring. The positive ion spectra of these acids
consist simply of a protonated molecular ion (see Figure
2) and display no loss of HF, confirming that this is a
true   mass   spectral   fragmentation   rather than chemical
thermolysis occurring in the   vaporizer.   Isotopic enrich-
ment of one carbon atom in the pyridine ring with ^{13}C produces the characteristic doublet ions in the mass spectra
of these metabolites.
      The monoacids in this series produce a weak M-H ion
at m/z 370 (m/z 371 for ^{13}C labeled monoacid) and abundant
ions from loss of HF (Figure 10). In contrast to the
diacids 2 and 8, monoacids undergo only one loss of HF,
and replacement of fluorine by hydrogen from the plasma
results in the weak ion at m/z 333. The characteristic
ion at m/z 391 arises from attachment of acetonitrile
radicals to preformed ions in the source. This is a process frequently observed during electron capture ionization (14). The effect of derivatization on the mass spectra is illustrated by the spectrum of the parent herbicide
(Figure 10, lower panel), which contains a strong molecular ion with the loss of a methyl radical as its only significant fragmentation.
      The sulfur atoms of dithiopyr have a pronounced role
in determining the metabolic fate of this fluorinated
pyridine. This point is illustrated by the metabolism of
another herbicide candidate in this class, which lacks
sulfur and is represented in Figure 11 by the generalized
structure 10. The pyridine 10 undergoes metabolism at a
much slower rate than dithiopyr and undergoes initial
oxidation of the isobutyl sidechain rather than the oxida-
tive ester hydrolysis observed with dithiopyr. The total
ion current trace from negative ion LC/MS analysis of 200
uL ofrat urine collected over a 24 hour period is shown
in Figure 11. The lactone 12 is the major component in
the urine, and presumably arises from hydroxylation of the
isobutyl sidechain followed by lactonization. Both isomers are observed, but the CF$_3$ lactone is much more abundant than the isomeric CF$_2$ lactone. We rationalize this
regiochemical preference in lactonization by the involvement of intramolecular hydrogen bonding between the CF$_2$H
group and the carboxyl group. Hydrolysis to the carboxylic acids 14 and 15 is a minor pathway, in contrast to
dithiopyr metabolism. Further oxidation of the isobutyl
sidechain produces the vinyl monoacid 13, as well as a
terminal carboxylic acid 16. These structural assignments were further corroborated by isolation, derivatization, and in some cases NMR analyses.
      The mass spectrum (Figure 12) of the major lactone
12 features deprotonation and loss of HF from the molecular ion. The much greater abundance of the M-H ion at m/z
338 relative to the monoacids of dithiopyr (Figures 7 and
10) may be due to the increased cyclization of this

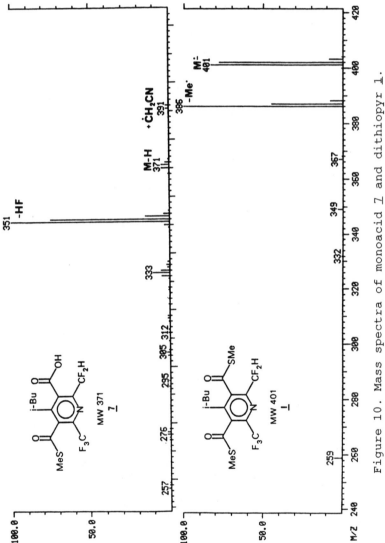

Figure 10. Mass spectra of monoacid 7 and dithiopyr 1.

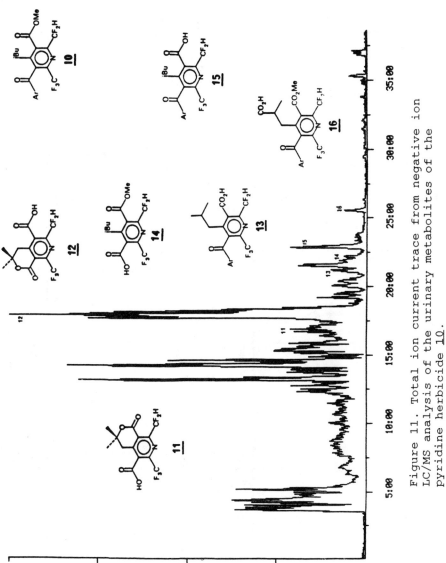

Figure 11. Total ion current trace from negative ion LC/MS analysis of the urinary metabolites of the pyridine herbicide 10.

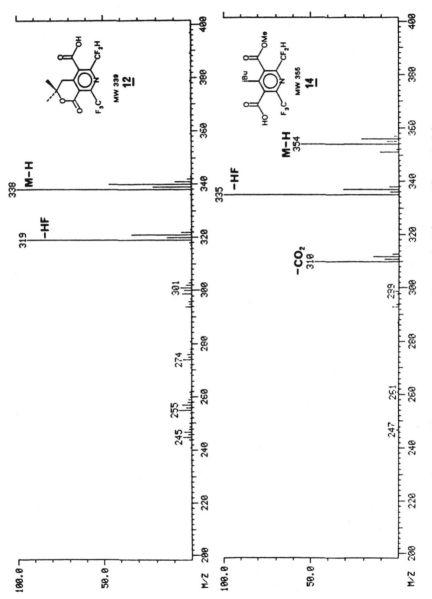

Figure 12. Mass spectra of metabolites 12 and 14.

species. The monoacid 14 produces a less abundant M-H ion which then undergoes a loss of carbon dioxide to form the ion at m/z 310. This loss of carbon dioxide is generally more pronounced for monoacids having the carboxylate group on the CF_3 side of the pyridine ring.

Hydrogen bonding between the CF_2H group and the carboxylate may promote lactone formation and inhibit decarboxylation in CF2 monoacids such as 15. The mass spectrum (Figure 13) of acid 15 displays a very weak doublet ion arising from decarboxylation at m/z 346, 347. In contrast to the other monoacids, 15 produces an abundance of molecular ions as well as the usual losses of hydrogen and HF. This may be due to its larger electron capture cross section since an aryl substituent is present in this compound. A similar effect is observed with the sidechain carboxylic acid 16 which also contains the aryl group but does not display loss of HF as an initial pathway. This is consistent with the hypothesis that the initial loss of HF results in the formation of a lactone adjacent to the CF_2H group.

Analysis of In Vivo Fecal Metabolites by LC/MS. Besides the analysis of urinary metabolites, animal metabolism studies of course require characterization of fecal metabolic products. We have found that rapid analysis of fecal extracts can also be accomplished by LC/MS, although more extensive cleanup is required. A general procedure for the extraction and LC/MS analysis of rat fecal metabolites is given in Figure 14. This procedure can easily be completed in an afternoon to provide a preliminary indication of metabolite structures. Based on this information the appropriate derivatives can be prepared for additional characterization, if necessary.

The selectivity of electron capture ionization for fluorinated compounds provides a convenient means to locate and identify fecal metabolites of dithiopyr. Figure 15 presents the total ion current trace obtained from negative ion LC/MS analysis of a rat fecal extract. In contrast to urinary samples, numerous peaks were observed in the fecal analysis which were associated with peptide conjugation. Most of the intermediates of the mercapturic acid pathway were found (e.g. 5, 17, 18, 19) along with two additional conjugates 20 and 21. Metabolites 20 and 21 are the CF2 and CF3 regioisomers of a thioglycolic acid conjugate which presumably arise from further degradation of the cysteinyl conjugate 18. As was observed in the urinary metabolite analyses, the isomeric monoacids 6 and 7 were present along with the diacid 2 and the isomeric lactones 11 and 12. No parent dithiopyr was observed in the aqueous portion of the fecal extract.

The PPINICI experiment again proved to be very useful in the analysis of fecal metabolites, by providing complementary information from both the positive and negative

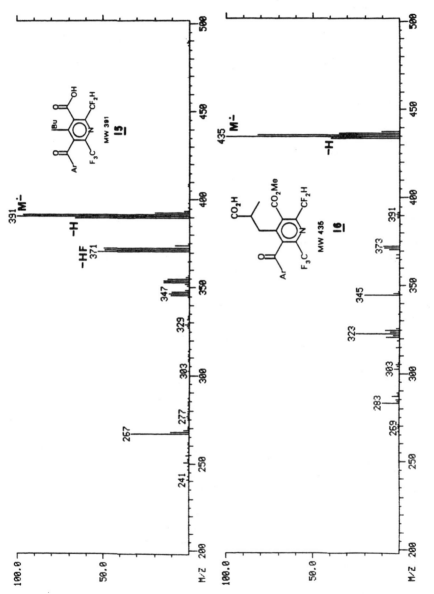

Figure 13. Mass spectra of metabolites 15 and 16.

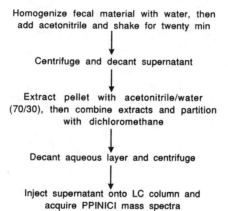

Figure 14. Flowchart for LC/MS analysis of fecal metabolites.

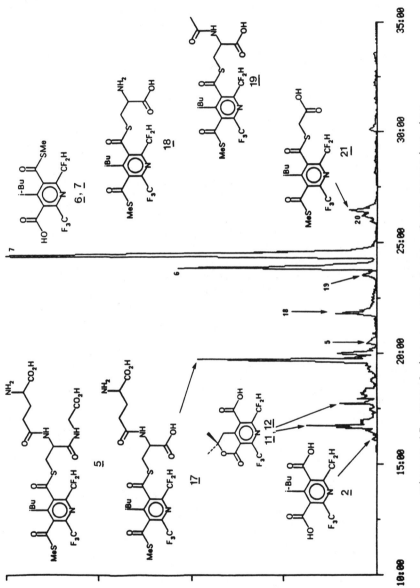

Figure 15. Total ion current trace from negative ion analysis of dithiopyr fecal metabolites.

ion analyses. As shown in Figure 16, chromatograms from the PPINICI experiment again illustrate the ionization selectivity characteristic of this class of compounds. The lower panel displays the negative ion trace from the PPINICI analysis and contains essentially only the peaks associated with fecal metabolites, whereas the positive ion trace (upper panel) contains numerous peaks associated with normal fecal constituents. The two monoacids 6 and 7 were the only pyridine-derived metabolites which could be identified from the positive ion chromatogram. The positive ion spectra were useful nonetheless, because they provided confirmation of molecular weights, and in some cases provided molecular weight information for peptide conjugates which underwent substantial fragmentation after electron capture.

The mass spectra of the cysteinyl conjugate 18 illustrate the complementary nature of positive and negative ion analysis (Figure 17). The protonated molecular ion is evident in the positive ion spectrum along with a major fragment ion at m/z 388 for a protonated thioacid. The ion at m/z 388 arises from loss of 2-aminoacrylic acid from the molecular ion. The anion of the thioacid is the highest mass ion observed in the negative ion spectrum, which provides little structural information otherwise. As mentioned earlier, cleavage of the peptide moiety to produce the anion at m/z 386 is a process which is characteristic of peptide conjugates in this class of compounds. It is useful to plot the ion intensities of m/z 386 and m/z 387 after negative ion analysis in order to ascertain which peaks in a chromatogram are associated with peptide conjugates. Once such conjugates are located, the corresponding positive ion spectrum generally provides molecular weight information for each individual conjugate.

The presence of an N-acetyl group in the cysteine moiety greatly increases the likelihood of obtaining molecular ions in both the positive and negative ion modes. Thus the mercapturic acid conjugate 19 provides perhaps the best spectra of any peptide conjugate. Figure 18 illustrates the molecular weight assignment of the mercapturic acid conjugate in both analysis modes. Further confirmation of the molecular weight is provided by the loss of HF from the molecular ions. The characteristic cleavage of the peptide moiety leading to the thioacid is evident in both spectra, as was the case for other peptide conjugates.

A type of peptide conjugate not frequently observed in mammalian metabolism is the glutamylcysteine conjugate 17. This conjugate is similar to the glutathione conjugate in that no molecular ions are observed in either positive or negative ion mass spectra. The highest-mass ion observed in the positive ion spectrum (Figure 19, upper panel) is at m/z 475 and corresponds to a protonated cysteinyl conjugate. Further fragmentation to the protonated thioacid leads to the most abundant ion at m/z 388.

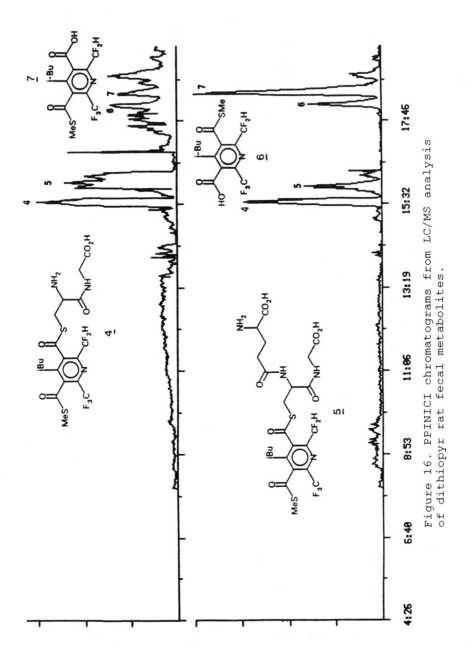

Figure 16. PPINICI chromatograms from LC/MS analysis of dithiopyr rat fecal metabolites.

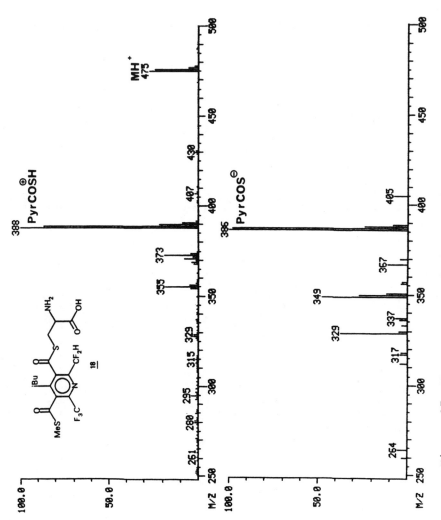

Figure 17. Mass spectra of the cysteinyl conjugate 18.

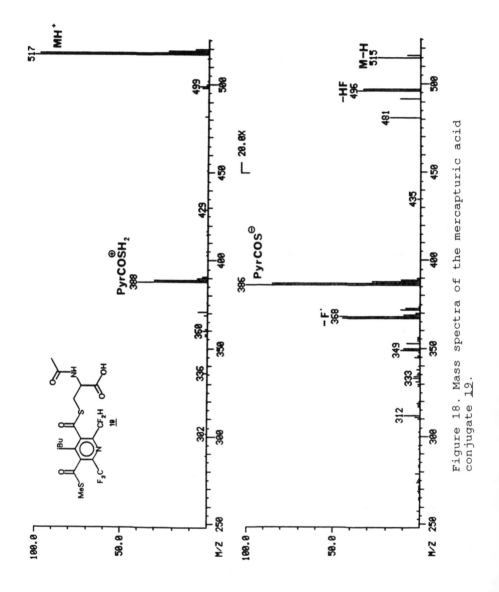

Figure 18. Mass spectra of the mercapturic acid conjugate <u>19</u>.

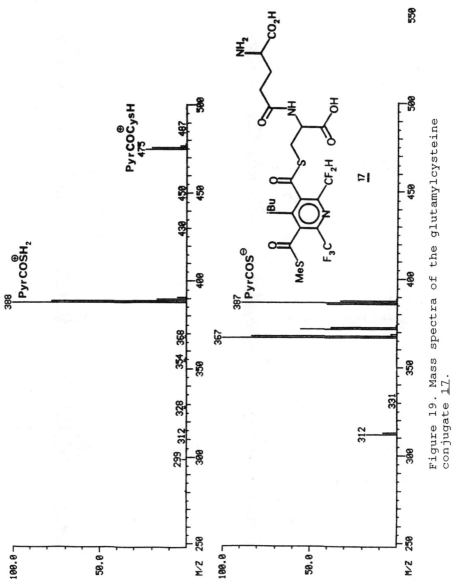

Figure 19. Mass spectra of the glutamylcysteine conjugate 17.

Figure 20. Fast atom bombardment negative ion mass spectrum of the glutamylcysteine conjugate 17.

By contrast, the negative ion spectrum merely indicated that this compound was a peptide conjugate since the highest-mass ion was at m/z 386.

Since the cysteinyl conjugate, 18, had already been identified as another peak in the chromatogram, the peak for compound 17 was isolated and methylated. The derivative gave a highest-mass ion at m/z 489, indicating the presence of a single carbolylate in this ion. The underivatized material was submitted to FAB analysis with the results shown in Figure 20. Negative ion FAB analysis was found to give the best sensitivity for this conjugate, and a M-H ion at m/z 602 was observed. Assignment of this doublet as the molecular anion was confirmed by sodium and potassium adducts observed at m/z 624 and m/z 640, respectively. The fragmentation pathway leading to the thioacid anion at m/z 386 dominates the FAB spectrum as it does in the LC/MS negative ion spectrum. The technique of FAB/MS complements negative ion LC/MS very well in that FAB analysis frequently provides molecular weights for peptide conjugates which undergo immediate fragmentation after electron capture. This same fragmentation provides a convenient method to locate quickly and sometimes identify such conjugates from a crude biological mixture by negative ion LC/MS.

Conclusion

In conclusion, the data presented here demonstrate that the technique of negative ion LC/MS can provide structural information on nanogram quantities of metabolites from crude biological samples with minimal purification. Most of the metabolites can be identified from mass spectra without derivatization, although corroboration of structural assignments by derivatization or FAB/MS is often desirable. However, the final analysis is greatly simplified by a knowledge of the derivatives needed and which ions to look for. Increased fragmentation associated with electron capture generally provides greater structural information which can be readily correlated with modifications in metabolite structure. In an era of ever-increasing demands on the sensitivity of analytical techniques, use of negative ion LC/MS can provide metabolism chemists with a simple yet tremendously powerful tool for metabolite structure identification.

Acknowledgments

The authors would like to thank Dr. Paul Feng, Mr. Jeff Duke, Ms. Sharon Moran, and Ms. Karen Chapman for providing metabolite samples for mass spectral analysis.

Literature Cited

1. Stemmler, E.M.; Hites, R.A.; Arbogast, B.; Budde,
    W.L.; Deinzer, M.L.; Dougherty, R.C.; Eichelberger,
    J.W.; Foltz, R.L.; Grimm, C.; Grimsrud, E.P.;
    Sakashita, C.; Sears, L.J. Anal. Chem. 1988, 60, 781.

2. Dougherty, R.C. Anal. Chem. 1981, 53, 625A.

3. Hunt, D.F.; Crow, F.W. Anal. Chem. 1978, 50, 1781.

4. Bruins, A.P. Advances in Mass Spectrometry 1985;
    John Wiley and Sons, 1986, p 119.

5. Hunt, D.F.; Crow, F.W. Anal. Chem. 1976, 48, 2098.

6. Stemmler, E.A.; Hites, R.A. Biomed. Environ. Mass
    Spectrom. 1988, 17, 311.

7. Harrison, A.G. Chemical Ionization Mass Spectrome-
    try; CRC Press, Boca Raton, 1983; p 27.

8. PPINICI is a trademark of Finnigan Corporation.

9. Kimmel, E.C.; Casida, J.E.; Ruzo, L.O. J. Agric.
    Food Chem. 1986, 34, 157.

10. Lehman, J.P.; Ferrin, L.; Fenselau, C.; Yost, G.S.
    Drug Metab. Dispos. 1988, 9, 15.

11. Bean, M.F.; Morrell, S.P.; Dulik, D.M.; Fenselau, C.
    Proc. 35th ASMS Conference on Mass Spectrometry and
    Allied Topics, 1987, p 1104.

12. Rashed, M.S.; Meyers, T.G.; Nelson, S.D. Proc. 35th
    ASMS Conference on Mass Spectrometry and Allied
    Topics, 1988, p 456.

13. Parker, C.E.; de Wit, J.S.M.; Smith, R.W.;
    Gopinathan, M.P.; Hernandez, O.; Tomer, K.B.;
    Vestal, C.H.; Sanders, J.M.; Bend, J.R. Biomed.
    Mass Spectrom. 1988, 17, 623.

14. Sears, L.J.; Campbell, J.A.; Grimsrud, E.P. Biomed.
    Mass Spectrom. 1987, 15, 401.

RECEIVED October 24, 1989

PHARMACEUTICALS AND METABOLISM

Chapter 8

Analysis of Xenobiotic Conjugates by Thermospray Liquid Chromatography/Mass Spectrometry

Deanne M. Dulik, George Y. Kuo, Margaret R. Davis, and Gerald R. Rhodes

Department of Drug Metabolism, Smith, Kline and French Research Laboratories, Swedeland, PA 19479

Positive ion thermospray LC/MS and LC/MS/MS analysis of xenobiotic conjugates obtained either from biological fluids or from enzymatic/chemical synthesis provides important information for the structure elucidation of this class of polar compounds. Conjugate metabolites amenable to thermospray LC/MS analysis include sulfate esters, glucuronides, taurine and carnitine conjugates, and mercapturic acid pathway conjugates. Thermospray ionization of these metabolites is best achieved in acidic buffers such as ammonium acetate; molecular ions are observed as $[M+H]^+$ or $[M+NH_4]^+$. Fragment ions are typically formed by loss of the conjugate moiety. Further structural information may be obtained through fragmentation of sample ions by collision-activated dissociation.

Hepatic metabolizing enzymes play a key role in the biotransformation of xenobiotics to polar, ionized species which are more readily excreted by various routes of elimination. The routes of hepatic biotransformation are designated generally as the Phase I reactions of functionalization (oxidation, reduction, and hydrolysis) and Phase II reactions of conjugation. The Phase II metabolites arise from conjugation with several substrates, including sulfonic acid, glucuronic acid, glutathione, amino acids, taurine, and carnitine. For

0097–6156/90/0420–0124$06.00/0

many years, the mass spectrometric analysis of this class of conjugates was hindered due to both their high polarity and thermal lability (1). More recently, LC/MS techniques, particularly by thermospray ionization mode, have proven to be successful for the structural characterization of conjugate metabolites from both *in vitro* and *in vivo* sources (2-8).

Positive ion thermospray LC/MS and LC/MS/MS methods carried out on a triple quadrupole mass spectrometer have been successfully utilized in our laboratory to elucidate the structures of metabolites of several compounds currently undergoing development as drug candidates. Samples were obtained from both *in vivo* sources (urine, bile, or plasma) and analyzed directly from the biological fluid or from *in vitro* enzymatic/chemical methods. In all cases, buffer ionization mode using ammonium acetate in the HPLC mobile phase was employed for ionization of the metabolites of interest.

MATERIALS AND METHODS

Chemicals

SK&F 96148, SK&F 86466, and SK&F 82526 were obtained from Drug Substances and Products, Smith Kline and French Laboratories, Swedeland, PA. ^{14}C-SK&F 96148 and ^{14}C-SK&F 86466 used for *in vivo* metabolism studies were provided by the Department of Radiochemistry, Smith Kline and French Laboratories, Swedeland, PA. Menadione (2-methyl-1,4-naphthoquinone) and its glutathione and N-acetyl cysteine conjugate were prepared by chemical synthesis under basic conditions and were kindly provided by Thomas Jones, Ph.D., Department of Pathology, University of Maryland School of Medicine. The regioisomeric phenolic glucuronides of SK&F 82526 were synthesized by reaction with rabbit liver microsomal uridine 5'-diphosphoglucuronyltransferases immobilized onto cyanogen bromide activated Sepharose 4B according to a published method (9). The products of the enzymatic incubation were purified using solid phase extraction before LC/MS analysis.

Animal Studies

Bile duct cannulated Sprague-Dawley rats (male and female) were administered ^{14}C-SK&F 96148 diluted with nonlabeled compound at a total dose of 100 mg/kg orally. Bile was collected at several time intervals up to 24 hrs post dose, acid stabilized with 10% glacial acetic acid and frozen immediately after collection. Female beagle

dogs were administered ^{14}C-SK&F 86466 diluted with
nonlabeled compound at a total dose of 10 mg/kg orally.
Urine samples were collected for 24 hrs and frozen
immediately.

LC/MS and LC/MS/MS Analysis

LC/MS and LC/MS/MS spectra were acquired on a triple
quadrupole mass spectrometer (TSQ-45, Finnigan MAT) using
a conventional Finnigan thermospray ion source. No
electron filament or discharge electrode was employed. In
general, ion source conditions for the polar conjugate
metabolites were: block temperature 200°-250°C; vaporizer
temperature 120°-130°C; repeller voltage 40-50 v. Spectra
were recorded in 1.95 second scans over the mass range
120-650 amu. LC/MS/MS analyses were performed with an
electron multiplier voltage of 1700, collision energy -25
to -30 eV, and argon collision gas pressure of 1.0 mtorr.
Reversed phase HPLC conditions were optimized for
each compound of interest. In all cases, the flow rate
was 1.1 ml/min, optimized for ionization in the
thermospray interface. A Kratos Spectroflow Model 783 UV
detector was placed in line before the mass spectrometer
and the eluent monitored at 237 or 254 nm for all
compounds. For SK&F 96148 biliary metabolites, the
following conditions were employed: Brownlee RP-300 (100
x 4.6 mm, 7 μm); solvent A, ammonium acetate (0.1 M)
adjusted to pH 5.3 with glacial acetic acid, solvent B,
acetonitrile; linear gradient conditions, time 0 min 25%
solvent B, time 25, 70% solvent B. This HPLC method was
also employed for characterization of the sulfate
conjugate of SK&F 86466 in dog urine and the N-
acetylcysteine conjugate of menadione, except that the
linear gradient conditions were 0-50% solvent B in 20 min.
A loop injection technique was employed for analysis of
the glutathione conjugate of menadione, with an RP-300
guard column (30 x 4.6 mm, 7 μm) and either 25% or 75%
isocratic solvent B. In the case of the glucuronide
conjugates of SK&F 82526, solvent B was methanol under 20%
isocratic conditions; the HPLC column was an Altex C_{18}
(150 x 4.6 mm, 5 μm).

RESULTS AND DISCUSSION

Biliary Metabolites of SK&F 96148

The total ion chromatogram for male rat bile (50 μl
direct injection) after oral dosing with ^{14}C-SK&F 96148 is
shown in Figure 1. Five drug-related components (as

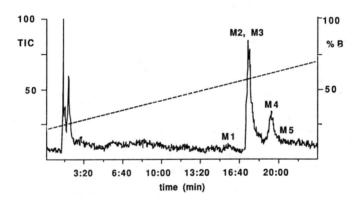

Figure 1.   Thermospray LC/MS total ion chromatogram
for male rat bile after oral dosing with SK&F 96148
(100 mg/kg).   Major drug related components are
designated M1-M5.   Linear LC gradient indicated by
dashed line (%B is percent acetonitrile).

confirmed by monitoring of ^{14}C by fraction collection and
liquid scintillation counting) were observed, designated
as M1-M5. The thermospray LC/MS and LC/MS/MS spectra for
compounds M1-M5 are shown in Figures 2-6.

Metabolite M1 gave a protonated molecular ion at m/z
477, 479 and an ammonium adduct at m/z 494, 496 (Fig. 2a).
The CAD daughter ion spectrum of m/z 494 showed diagnostic
fragment ions for loss of the taurine moiety at m/z 126
and loss of the teminal carboxyl group from the dehydrated
parent structure at m/z 306 (Fig. 2b). These observations
(along with proton NMR decoupling data, not shown) were
consistent with the proposed structure of M1 as the
taurine conjugate of a hydroxylated metabolite of the
parent (M5). In addition, the assigned position of
hydroxylation was supported by dehydration in the MS/MS
mode to give a strong fragment ion at m/z 459 and further
decarboxylation of the parent structure to give a fragment
at m/z 334.

Metabolite M2 gave a protonated molecular ion at m/z
461, 463 and a base peak ammonium adduct at m/z 478, 480
(Fig. 3a). The minor fragment ion at m/z 387 may arise
from cleavage of the ethanesulfonic acid moiety; the ion
observed at m/z 533 is most likely the ammonium adduct of
the endogenous bile acid, taurocholic acid. The CAD
daughter ion spectrum of m/z 478 produced fragments at m/z
126 corresponding to the loss of taurine and at m/z 308
due to subsequent decarboxylation of the parent (Fig. 3b).
This information is consistent with the assignment of M2
as the taurine conjugate of the parent.

Metabolite M3 was the major biliary metabolite of
SK&F 96148. Thermospray LC/MS analysis of M3 showed an
ammonium adduct at m/z 547, 549 and a prominent fragment
ion for the aglycone parent at m/z 371, 373 (Fig. 4a).
The ion at m/z 413 corresponds to thermal acetylation of
the parent after glucuronide hydrolysis in the ion source
and is observed for other glucuronides. The CAD daughter
ion spectrum of m/z 547 produced fragment ions
corresponding to a loss of glucuronic acid (m/z 354, 356,
and 371, 373), a base peak fragment due to decarboxylation
of the parent structure at m/z 308, and other minor
fragments of the parent molecule (Fig. 4b). These data
supported the structure assignment of M3 as the acyl
glucuronide conjugate of the parent. Subsequent proton
NMR analysis (not shown) confirmed the structure as the 1-
O-acyl glucuronide.

Metabolite M4 was analyzed by thermospray LC/MS
(Fig. 5a) and produced a protonated molecular ion at m/z
497, 499 (confirmed by fast atom bombardment mass spectral
analysis). This mass spectrum, as well as proton NMR
analysis (not shown) was consistent with the acyl
carnitine conjugate of the parent. Fragmentation occurred
via loss of the carnitine moiety to give m/z 371, 373; the
fragment ion at m/z 455, 457 could correspond to an

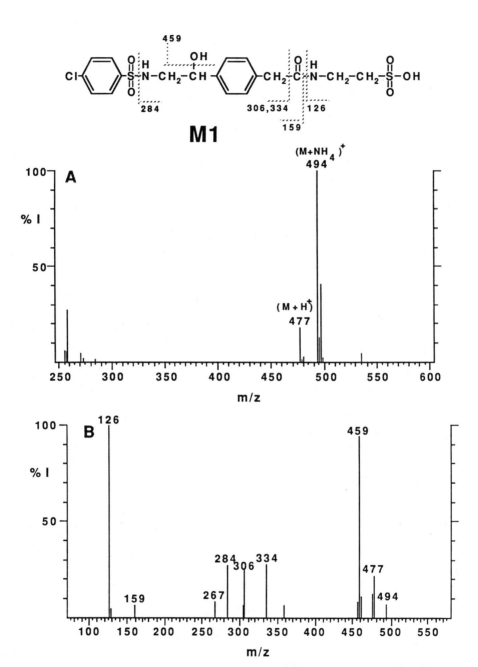

Figure 2. Thermospray A) LC/MS and B) LC/MS/MS spectra of SK&F 96148 biliary metabolite M1, the hydroxylated taurine conjugate.

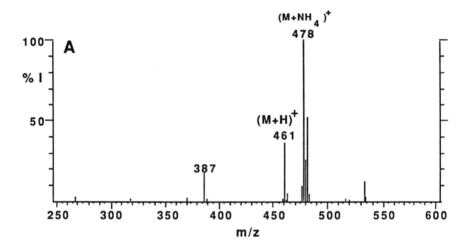

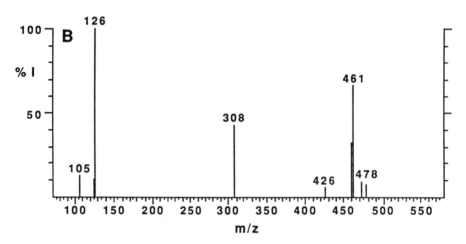

Figure 3.   Thermospray A) LC/MS and B) LC/MS/MS
spectra of SK&F 96148 biliary metabolite M2, the
taurine conjugate.

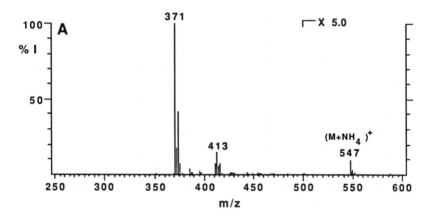

M3

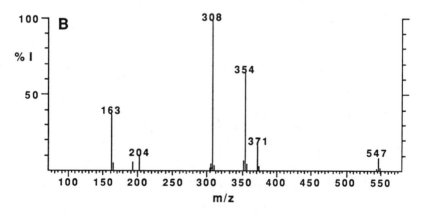

Figure 4. Thermospray A) LC/MS and B) LC/MS/MS spectra of SK&F 96148 biliary metabolite M3, the 1-0-acyl glucuronide conjugate.

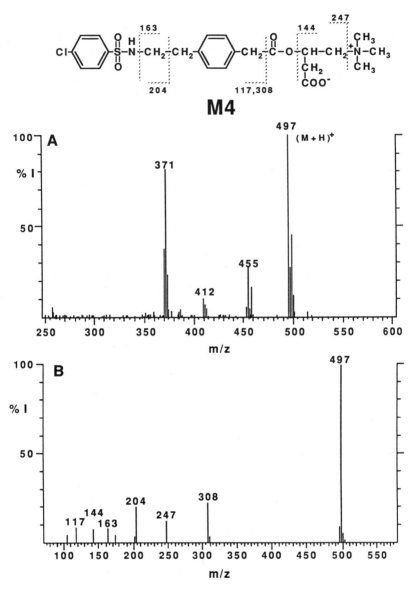

Figure 5.   Thermospray A) LC/MS and B) LC/MS/MS
spectra of SK&F 96148 biliary metabolite M4, the acyl
carnitine conjugate.

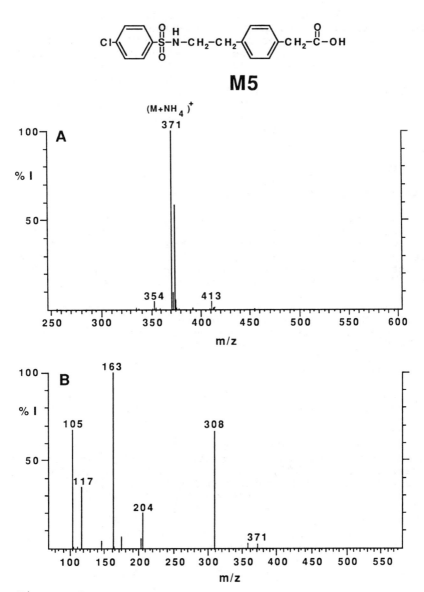

Figure 6.   Thermospray A) LC/MS and B) LC/MS/MS spectra of parent SK&F 96148 (designated as M5).

ammonium adduct after loss of trimethylamine. Very little
fragmentation of this metabolite occurred under CAD
conditions (Fig. 5b). Small fragments at m/z 308, 204,
and 163 arise from the parent; the fragment at m/z 247
could occur from loss of trimethylamine.

M5 was shown to be parent SK&F 96148 by coelution in
the HPLC with authentic standard and by its LC/MS analysis
(Fig. 6a). Thermospray LC/MS analysis gave an intense
ammonium adduct at m/z 371, 373 and a weak protonated
molecular ion at m/z 354, 356. CAD daughter ion spectrum
at m/z 371 showed fragment ions at m/z 308 corresponding
to decarboxylation, m/z 204 corresponding to cleavage of
the ethylene bridge, and m/z 163 corresponding to loss of
chlorobenzenesulfonamide (Fig. 6b). Ions at m/z 105 and
117 arise from further cleavage of these initial
fragments.

Urinary Sulfate Metabolite of SK&F 86466

The thermospray LC/MS analysis of a major urinary
sulfate ester metabolite of SK&F 86466 is shown in Figure
7. A weak ammonium adduct was observed at m/z 295, 297
with an isotope distribution consistent with the proposed
structure. The major fragment ion at m/z 182, 184 occurs
from loss of the sulfate ester moiety. The position of
sulfation at either C7 or C9 of the molecule was
determined by proton NMR NOE difference studies (data not
shown). No protonated molecular ion was observed for this
sulfate ester; this would be expected, based on the proton
affinity of the highly acidic sulfate ester moiety.

Glucuronide Metabolite of SK&F 82526

The thermospray LC/MS spectrum of the phenolic
glucuronide conjugates of SK&F 82526 is shown in Figure 8.
Two regioisomeric monoglucuronides were synthesized using
immobilized microsomal UDP-glucuronyltransferases from
rabbit liver and separated by LC/MS, producing identical
thermospray spectra. A protonated molecular ion was
observed at m/z 482, 484. A large fragment ion was
observed at m/z 306, corresponding to loss of glucuronic
acid. The ion at m/z 194 corresponds to the ammonium
adduct of glucuronic acid. In addition, the ion at m/z
272 may occur via contraction of the benzazepine ring to a
more stable 6-membered ring; this fragment is also
observed in the parent molecule.

Mercapturic Acid Pathway Metabolites of Menadione

The mercapturic acid pathway metabolites of
menadione, the 3-glutathionyl and corresponding N-
acetylcysteine conjugate (mercapturic acid) are a result
of a detoxicative pathway designed to remove reactive
electrophilic substrates from mammalian systems (10). The

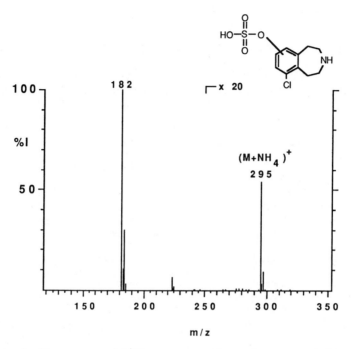

Figure 7. Thermospray LC/MS spectrum of a urinary metabolite of SK&F 86466, the sulfate ester conjugate.

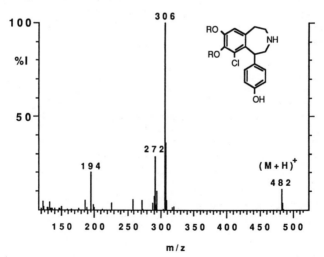

Figure 8.   Thermospray LC/MS spectrum of the phenol monoglucuro-nide conjugates of SK&F 82526 obtained   by   immobilized   enzyme synthesis.

LC/MS analysis of the N-acetylcysteine conjugate is shown
in Figure 9. The base peak molecular ion was observed at
m/z 334. Very little fragmentation occurred with the
exception of a small amount of S-C cleavage to produce a
protonated thiol at m/z 202.

The glutathione conjugate (γ-glutamylcysteinyl-
glycine, designated as SG) of menadione is the first
product in the mercapturic acid pathway. The thermospray
LC/MS spectrum of this conjugate using 25% acetonitrile :
75% 0.1 M ammonium acetate, pH 5.3 as the mobile phase is
shown in Figure 10a. Extensive thermal degradation was
observed under these conditions, with no molecular ion for
parent (molecular weight 477) and a base peak fragment ion
at m/z 202 corresponding to S-C bond cleavage. However,
upon increasing the acetonitrile composition in the mobile
phase to 75% and decreasing the vaporizer temperature from
125°C to 118°C, the LC/MS spectrum in Figure 10b was
produced. In this case, a protonated molecular ion was
observed at m/z 478 with a sodium adduct at m/z 500 (most
likely from the synthetic method employed). The fragment
ions at m/z 129 and 147 correspond to protonated glutamic
acid and its corresponding ammonium adduct. Other minor
fragment ions correspond to cleavage of the tripeptide and
C-S bond cleavage in the parent structure. It appears
that for thermally labile conjugates such as glutathione
conjugates, an increase in the percent organic modifier
and subsequent decrease in vaporizer temperature for
thermospray ionization results in improved molecular ion
production. This has been reported by Bean and coworkers
for similar compounds (11) and probably is a result of
decreased thermal hydrolysis in the presence of smaller
percentages of aqueous buffer.

CONCLUSIONS

Positive ion thermospray LC/MS and LC/MS/MS analysis
may be used to characterize a variety of polar, highly
ionized xenobiotic conjugates directly from biological
fluids or from in vitro synthetic methods. Several types
of conjugates, including sulfate esters, glucuronides,
conjugates with amino acids, taurine, and carnitine, and
mercapturic acid pathway metabolites may be analyzed using
this technique. Buffer ionization mode using ammonium
acetate is useful for analyses of this type. Compounds
which are highly thermally labile are more successfully
characterized with increased percentages of organic
modifier in the mobile phase and at reduced vaporizer
temperatures. LC/MS and LC/MS/MS analysis of xenobiotic
conjugates provides predictable fragmentation, primarily
via loss of the conjugate moiety. This fragmentation
combined with neutral loss analysis in the MS/MS mode
makes possible the screening for unknown metabolites of
xenobiotics.

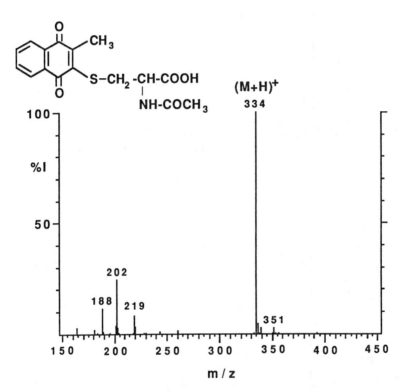

Figure 9. Thermospray LC/MS spectrum of 2-methyl(N-acetylcysteinyl)naphthoquinone (mercapturic acid of menadione).

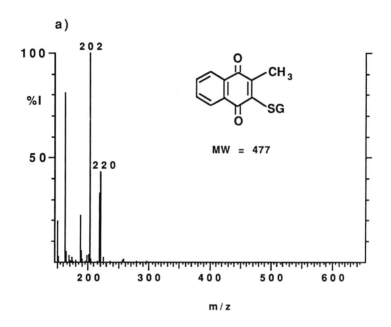

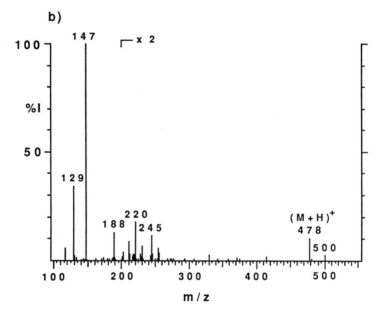

Figure 10.   Thermospray LC/MS spectrum of (3-glutathionyl)-2-methylnaphthoquinone using a) 25% acetonitrile as organic modifier, and b) 75% acetonitrile as organic modifier.

ACKNOWLEDGMENTS

The authors gratefully acknowledge M. Carbonaro, L. Gutzait, G. Joseph, and J. Kao, Department of Drug Metabolism, SK&F Research Labs, for providing samples from animal studies. We also thank T. Jones, Department of Pathology, University of Maryland School of Medicine, for providing authentic samples of the menadione conjugates. The assistance of C. DeBrosse, A. Goalwin, and W. Schaefer, SK&F Research Labs, is greatly appreciated.

LITERATURE CITED

1. Fenselau, C.; Johnson, L.P. Drug Metab. Dispos. 1980, 8, 274-283.
2. Liberato, D.J.; Fenselau, C.; Vestal, M.L.; Yergey, A.L. Anal. Chem. 1983, 55, 1741-1744.
3. Straub, K.M.; Rudewicz, P.; Garvie, C. Xenobiotica 1987, 17, 413-422.
4. Blake, T.J. A. J. Chromatogr. 1987, 394, 171-181.
5. Rudewicz, P; Straub, K.M. Drug Metab. Mol. Man [Eur. Drug Metab. Workshop, 10th] 1987, 208-212.
6. Liberato, D.J.; Millington, D.S.; Yergey, A.L. Anal. Chem. Symp. Ser. 1985, 24, 333-348.
7. Weidolf, L.O.G.; Lee, E.D.; Henion, J.D. Biomed. Environ. Mass Spectrom. 1988, 15, 283-289.
8. Parker, C.E.; DeWitt, J.S.; Smith, R.W.; Gopinathan, M.B.; Hernandez, O.; Tomer, K.B.; Vestal, C.H.; Sanders, J.M.; Bend, J.R. Biomed. Environ. Mass Spectrom. 1988, 15, 623-634.
9. Dulik, D.M.; Fenselau, C. Drug Metab. Dispos. 1987, 15, 473-477.
10. Chausseaud, L. Drug Metab. Rev. 1973, 2, 185-202.
11. Bean, M.F.; Pallante-Morell, S.L.; Dulik, D.M.; Fenselau, C. Anal. Chem., 1989, in press.

RECEIVED October 6, 1989

Chapter 9

Qualitative Analysis of Pharmaceuticals by Thermospray Liquid Chromatography/Mass Spectrometry

Nemadectins and Tetracyclines

Grant B. Kenion, Guy T. Carter, Jaweed Ashraf, Marshall M. Siegel, and Donald B. Borders

Medical Research Division, American Cyanamid Company, Pearl River, NY 10965

Our primary use of thermospray liquid chromatography-mass spectrometry (LC/MS) has been as a tool with which to search for novel compounds in fermentation broths. Two classes of compounds for which this technique has been particularly useful are the nemadectins (a family of antiparasitic macrolides recently isolated from Streptomyces cyaneogriseus sp. noncyanogenus (1)) and the tetracyclines. The optimal thermospray LC/MS compatible HPLC separation for the nemadectins alpha, beta, gamma, and lambda consists of a methanol/ water solvent gradient solvent system and a C-18 column. Interpretation of the thermospray mass spectra in positive and negative ion modes using either discharge electrode or filament ionization demonstrate that the electron capture ionization dominated negative ion spectra provide excellent molecular weight information, while the positive ion spectra provide useful structural information due to extensive fragmentation. For the tetracyclines, a thermospray compatible HPLC method using a C-4 column and a solvent system consisting of dimethylformamide and ammonium acetate buffer is capable of separating tetracycline, 6-demethyltetracycline, declomycin, and chlortetracycline. The positive ion thermospray mass spectra of the tetracyclines are simple, with the base peak consisting of the proton adduct of the parent compound. The negative ion thermospray spectra (with the discharge electrode on) display intense, structurally diagnostic fragmentation.

0097–6156/90/0420–0140$07.50/0

Nemadectins alpha, beta, gamma, and lambda (1-4) are produced via a fermentation process (2) that also produces a number of novel minor components of similar structure. Our initial objective was to adapt our HPLC method for thermospray LC/MS analysis. The second objective was to evaluate ionization and detection modes to determine the best combination in terms of sensitivity, molecular weight information, and structurally diagnostic fragmentation data. The ultimate goal was to develop an analytical method retaining the resolving power of the original HPLC method (1), with adequate sensitivity to detect minor components and sufficient fragmentation to differentiate isomeric compounds.

Thermospray mass spectra often provide excellent molecular weight information, with some additional structural information being provided by fragment ions, via an ionization mechanism similar in "softness" to chemical ionization (CI) (3). Spectra obtained by previous researchers from 22,23 dihydroavermectin B_{1a}, a macrolide similar in structure to the nemadectins, demonstrated that chemical ionization mass spectrometry(4) and thermospray LC/MS (5) provided excellent results. We anticipated, therefore, that the nemadectins would also be amenable to thermospray LC/MS analysis.

High performance liquid chromatography of tetracyclines using reversed phase columns (6-15) has been a subject of intense interest over the last decade and a half. Previous methods have generally used sodium phosphate (6), ammonium carbonate (13), citric acid(14), tetraalkyl ammonium salts (15), or other non-volatile buffering agents to improve resolution. Unfortunately, while these buffers are fine when using UV detection, they are not the buffers of choice for thermospray LC/MS analysis. We present a reversed phase HPLC separation that is thermospray compatible, as shown by positive and negative ion spectra and a total ion chromatogram showing good resolution between 6-demethyltetracycline, tetracycline, declomycin, and chlortetracycline. Our choice of dimethylformamide as the organic modifier was based on excellent solubility and chromatographic characteristics for tetracyclines obtained by R. Leese and F. Barbatschi (16) for an HPLC system using a non-volatile buffer. Ammonium acetate was chosen on the basis of its optimal (17) compatibility with thermospray LC/MS analysis. Initial attempts using perfluorinated buffers were plagued by high mass noise (probably from column bleed and cluter formation). Initial attempts using ammonium acetate and ammonium formate buffers with C-8 and C-18 columns did not provide adequate separation. A

solution of the four standards (6-demethyltetracycline,
tetracycline, declomycin, and chlortetracycline) eluted
as a single peak with very little retention. A C-4
reversed phase column was the only column we tested that
could provide adequate separation using ammonium acetate
or ammonium formate buffers.

MATERIALS AND METHODS

Nemadectins

   All mass spectral analyses other than positive ion
desorption chemical ionization (DCI) mass spectra were
performed using a Finnigan (San Jose, CA) TSP-46
Dedicated Thermospray LC/MS single quadrupole mass
spectrometer. Data were acquired using Revision 5.5 of
the INCOS data system software. The vaporizer temperature
for LC/MS analysis was set at 95°C, while the jet
temperature was set at 220°C. Optimization of temperature
parameters was performed using maximum molecular ion
intensity, gaussian peak shape, and maximum total ion
response as quality indicators. A mass range of 400-650
amu was scanned every 2 seconds. The electron multiplier
was set at 1800 volts. When in use, the discharge
electrode was set at 1 KV. The chemical ionization
experiments were performed using the EI/CI source. Ultra
Pure Grade methane purchased from Linde (Danbury, CT)
was used as the reagent gas. The ionizer temperature was
set at 100°C. Source pressure was optimized at 1.44 torr.
The electron energy was set at 70 eV. The electron
multiplier was set at 1300 volts. A mass range of 100-800
amu was scanned every 2 seconds. Samples were introduced
via a direct insertion probe heated from 30°C to 400°C at
120°C / min. (Structure 1)
   All positive ion DCI mass spectra were obtained from
a Finnigan MAT 90 double focusing magnetic sector mass
spectrometer. The electron energy was set at 150 eV. The
instrument was set to scan from 100-850 amu at 2 seconds
per decade with a 1 second interscan time. Ammonia was
used as the reagent gas at a source pressure of
1×10^{-4}Torr.
   A Spectra-Physics (San Jose, CA) SP 8800 Ternary
Gradient HPLC pumping system was used. Methanol and water
of HPLC grade were purchased from J.T. Baker (Phillipson,
NJ) and used without further purification. Injection of
samples for LC/MS analysis was performed using a Waters
(Springfield, MA) WISP Autoinjector.
   HPLC separations were accomplished with a Perkin
Elmer 3X3 CR C-18 column. A solvent gradient of 60%
methanol, 40% water to 90% methanol, 10% water in 30

minutes at a flow rate of 1.2 ml/min provided best
results. A total ion chromatogram from an LC/MS analysis
is shown in Figure 1 for a standard solution of (in
order of elution) nemadectins beta, gamma, alpha, and
lambda with a concentration of approximately 1 mg/ml of
each analyte. Use of ammonium acetate buffer causes gamma
and alpha to coelute, as does using acetonitrile instead
of methanol as the organic modifier. Authentic samples of
the nemadectins alpha, beta, gamma, and lambda were
obtained as previously described (1).

Tetracyclines

     All tetracycline, 6-demethyltetracycline,
declomycin, and chlortetracycline standards were provided
by Lederle Laboratories as the hydrochloride salts. All
fast atom bombardment spectra were obtained with a VG
(Manchester, UK) ZAB SE equipped with a cesium ion gun
operated at 30 kV. The optimum matrix was "magic bullet"
with a methanol solvent. (Structure 2)
     All thermospray mass spectra were obtained from a
Finnigan TSP-46 Dedicated Thermospray LC/MS coupled to a
Spectra-Physics SP-8800 Ternary HPLC system. A Rainin
Dynamax 150°A 25cm 12mm C-4 column (catalog #83-502-C)
with matching C-4 guard colum (catalog # 83-502G) was
used throughout. The 12mm particle size is important, as
pressure problems can arise when using dimethylformamide
as the organic modifier due to the high viscosity of the
solvent.
     The following solvent gradient system provides optimum
results at a flow rate of 1.2 ml/min:

TIME (min)	%0.1 M NH4OAC (2%ACN, pH 6.5)	%H2O%	DMF
0	80	15	5
10	80	0	20
35	80	0	20
40	50	0	50
60	50	0	50

     Although all of the standards presented in this
manuscript will elute at a solvent composition of 20%
dimethylformamide, other members of this class of
compounds will not elute with a solvent composition less
than 50% dimethylformamide.
     The optimum mass spectrometer parameters are as
follows:
     Scan Range: 300-800 amu, 2 seconds scan rate
     Vaporizer Temperature: 120°C, change to 130°C when
     DMF composition is above 20%.
     Jet Temperature: 250°C
     Discharge Electrode: 1KV

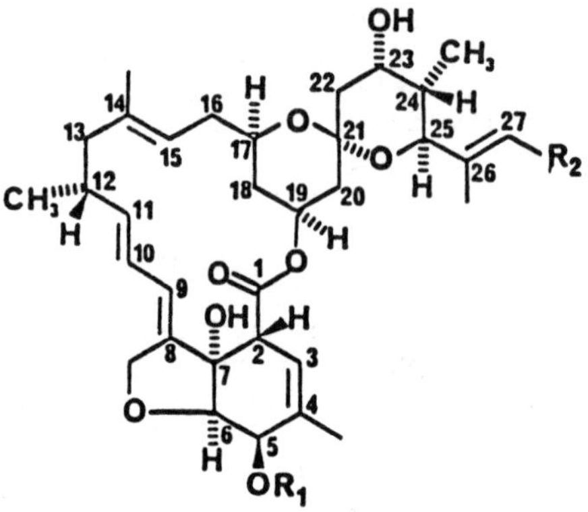

		R_1	R_2
LLF28249α	1	H	ipr
LLF28249β	2	H	Me
LLF28249γ	3	Me	Me
LLF28249λ	4	Me	ipr

Structure 1

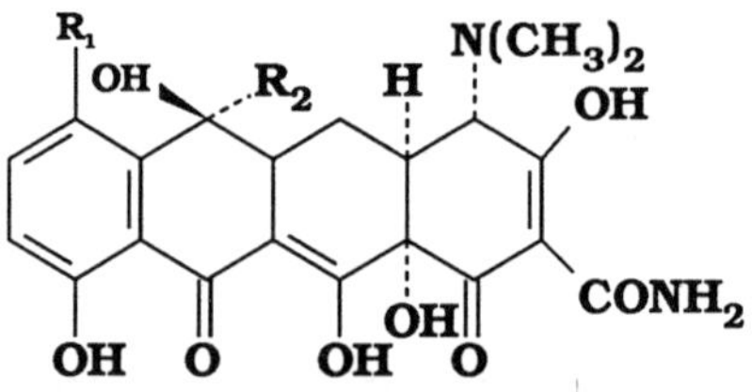

6-Demethyltetracycline	R_1=H, R_2=H
Tetracycline	R_1=H, R_2=CH$_3$
Declomycin	R_1=Cl, R_2=H
Chlortetracycline	R_1=Cl, R_2=CH$_3$

Structure 2

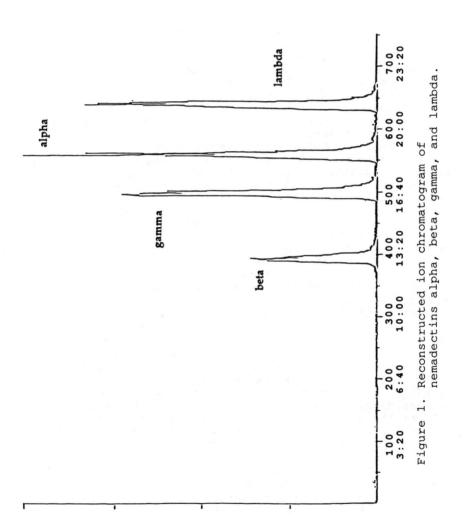

Figure 1. Reconstructed ion chromatogram of nemadectins alpha, beta, gamma, and lambda.

It should be noted that a high sample throughput using this analysis will often cause the quadruple rods to become contaminated. When using this method, one should check the raw data periodically and ensure that high mass assignment is stable.

RESULTS AND DISCUSSION

Nemadectins

The optimal HPLC separation conditions for the nemadectins precluded the use of "buffer only" ionization. The Finnigan TSP-46 thermospray source is equipped with a filament and a discharge electrode as choices for an external ionization method. Both positive and negative ion detection are available. The best choice for an ionization method and for the mode of ion detection would have to provide adequate sensitivity for minor component analysis, unambiguous molecular weight information, and sufficient fragmentation to differentiate between components with similar retention times and identical molecular weights.

Each of the standard nemadectins was analyzed using filament ionization in both positive and negative ion detection modes and using the discharge electrode in both detection modes. These data were then compared with regard to sensitivity, degree of fragmentation (and the resulting structural information content) , and quality of molecular weight information.

The positive ion spectra obtained using filament ionization (as shown in Figure 2 and in Table I) are characterized by an $(M+H^+)$ ion with a relative abundance of 10-20%. The most abundant fragment ions observed in these spectra correspond to consecutive losses of water from $(M+H)^+$. The spectra of the gamma and lambda components also contain ions resulting from loss of methanol (m/z 531 and 567; m/z 577 and 595 respectively), presumably arising from the C-5 methoxy group. Ions resulting from loss of three water molecules from gamma and lambda must result from hydrolysis at the acetal at C-21 or the ether at C-6 and subsequent loss of water, as the intact molecules only have two hydroxy groups each.

In addition to these eliminations, there are ions derived from retro Diels-Alder fragmentations in the spectra. Each of the compounds shows an ion derived from the retro Diels-Alder fragmentation of the $(M-2\ H_2O)^+$ species, fragmentation a in Scheme 1. The masses of these ions are useful in determining changes in the substitution pattern on the core part of the molecule apart from the C-25 side chain. Thus both alpha and beta give rise to m/z 465 ions, whereas gamma and lambda yield

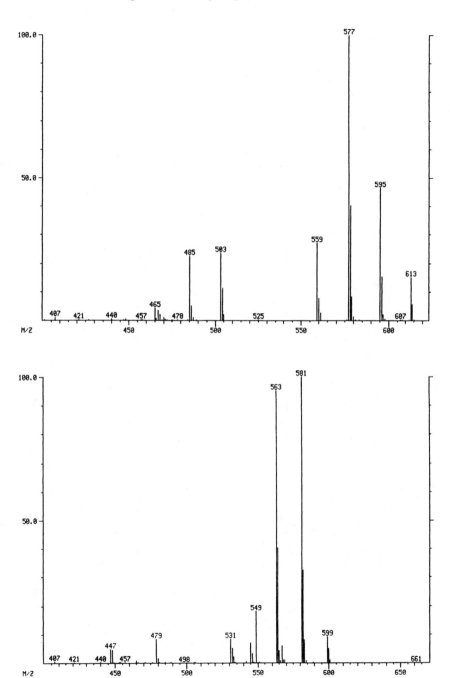

Figure 2. Positive ion spectra using filament ionization. Continued on next page.

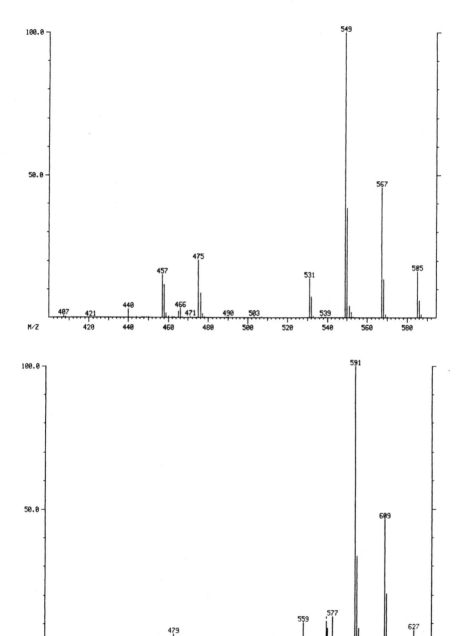

Figure 2. Continued.

Table I. Positive ion spectra [c]

	Alpha	Beta	Gamma	Lambda
$M+Na^+$	m/z 635(0,8,0)	m/z 607(0,5,0)	m/z 621(0,5,0)	m/z649(0,5,0)
$M+NH_4^+$	m/z 630(0,0,32)	m/z 602(0,0,23)	m/z 616(0,0,20)	m/z644(0,0,24)
$M+H^+$	m/z 613(20,10,60)	m/z 585(20,5,70)	m/z 599(21,5,48)	m/z627(12,5,45)
$M+H^+$ $-H_2O$	m/z 595(50,55,100)	m/z 567(45,43,100)	m/z 581(100,78,100)	m/z609(45,45,100)
$M+H^+$ $-2H_2O$	m/z 577(100,100,60)	m/z 549(100,100,63)	m/z 563(98,100,75)	m/z591(100,100,93)
$M+H^+$ $-H_2O$ $-MeOH$	not present	not present	m/z 549(12,25,10)	m/z 577(8,15,10)
$M+H^+$ $-3H_2O$	m/z 559(30,50,18)	m/z 531(20,37,17)	m/z 545(12,35,15)	m/z 573(18,50,20)
Retro Diels-Alder b	m/z 503(22,10,0)	m/z 475(30,5,0)	not present	not present
$M+H^+$ $-2H_2O-MeOH$	not present	not present	m/z 531(8,22,10)	m/z 559(5,15,10)
$M+H$ $-110-H_2O$	m/z 485(20,20,0)	m/z 457(15,15,0)	not present	not present
Retro Diels-Alder a	m/z465(12,20,10)	m/z 465(10,15,15)	m/z 479(12,12,10)	m/z 479(10,12,10)

c. Ions are listed as m/z xxx(A,B,C) where A, B, and C are the relative intensities of the ions in spectra taken under filament ionization thermospray, discharge electrode ionization thermospray, and ammonia desorption chemical ionization conditions (respectively).

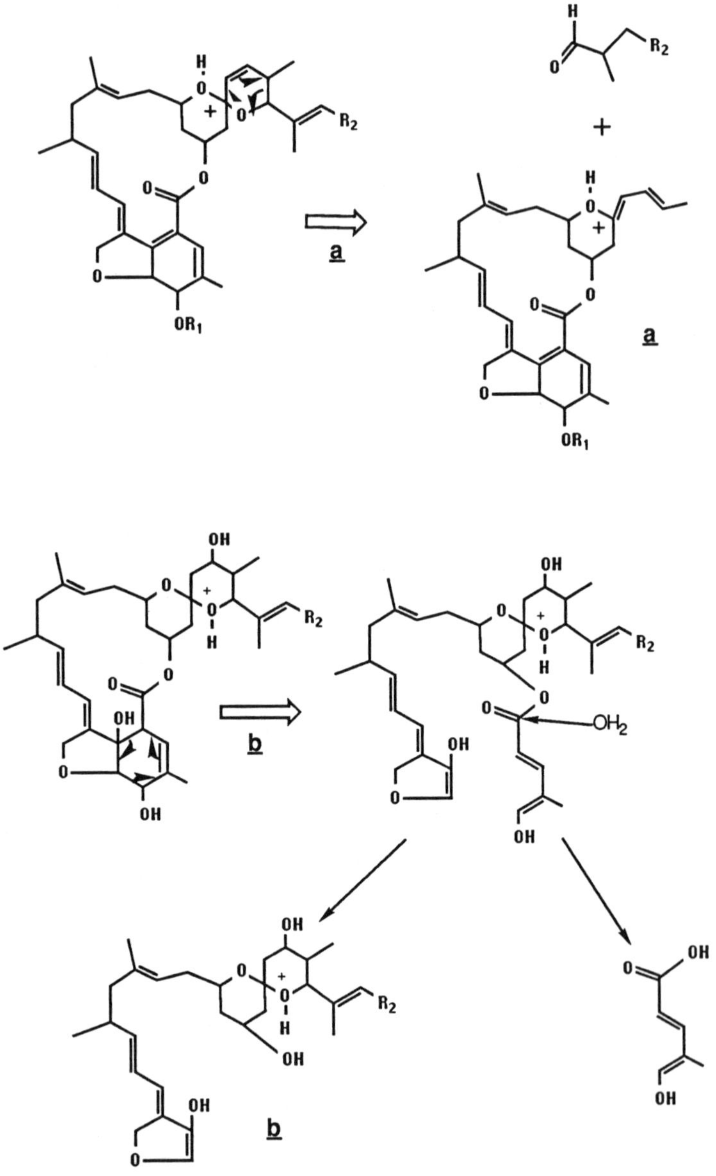

Scheme 1. Retro Diels-Alder fragmentations in
           positive ion spectra.

ions of m/z 479 due to the methyl substituent at O-5. The presence or absence of this methyl group on O-5 appears to be the deciding factor as to whether or not the second retro Diels-Alder fragmentation (b) shown in Scheme 1 occurs. This process yields relatively abundant ions for alpha (m/z 503) and beta (m/z 475), which contain an OH group at C-5, but no corresponding fragments were observed in the spectra of the 5-methoxy analogs, gamma and lambda. Similar behaviour for fragmentation b is evident in the negative ion thermospray spectra and the negative ion CI spectra , although these ions are have a low relative abundance in the chemical ionization spectra. Apparantly, the "softer" ionization methods, such as thermospray and chemical ionization, provide enough energy to induce the retro- Diels-Alder fragmentation b only when a hydroxy group is at C-5. The methoxy group must increase the activation energy of the reaction, via electronic or steric effects, such that it is not favored over other ion forming processes in thermospray or chemical ionization, but that the barrier is not significant in the higher energy electron impact ionization spectra.

Positive ion spectra obtained via discharge electrode ionization (Table I) exhibit increased abundances of ions due to elimination of water and methanol at the expense of the molecular adduct ions. Other than an (M + Na^+) (approximately equal in abundance to the ($M+H^+$) ion) formed using discharge electrode ionization, the same ion masses are produced for both filament and discharge electrode ionization. Because of the greater number of ions at m/z greater than the masses of the analytes, the assignment of molecular weights would be more ambiguous than for the filament ionization spectra.The benefit of using filament ionization over discharge electrode ionization lies, therefore, in the apparently milder ionization and lesser background noise. Sensitivity is adequate, with a full spectrum obtainable from 265 ng of analyte.

Positive ion DCI spectra obtained using ammonia reagent gas produce ion masses similar to the positive ion thermospray spectra, with the exception of some lower mass fragment ions (not listed) of low relative intensity also found in the electron impact spectra (1), the ($M+NH_4$)$^+$ cluster ions, and the lack of fragment ions for alpha and beta from fragmentation pathway b. The basicity of the ammonia enhances dehydration, resulting in loss of the hydroxy group at C-7, such that pathway a is allowed but pathway b is not. The chemical ionization data have a higher relative proportion of molecular ion species to fragment species compared to the thermospray spectra due to the fact that they were obtained on a magnetic sector instrument and because the higher thermal energy

imparted to the analyte under filament or discharge
electrode ionization thermospray conditions enhances
fragmentation.

Negative ion spectra from filament ionization (as
shown in Figure 3 and Table II) are dominated by intense
M⁻ ions produced by electron capture. Ions resulting from
losses of water or methanol are greatly reduced in
abundance, relative to the corresponding positive ion
spectra, and there is no indication of type a retro
Diels-Alder fragmentation. Notably, the type b retro
Diels-Alder process still produces fragmentation ions in
the negative ion mode for the congeners with C-5 OH
substitution, yielding the corresponding M-110 amu ions.
The discharge electrode ionization spectra (not shown)
are virtually identical to the filament ionization
spectra, although the relative intensities of the
fragment ions may be up to 30 % higher. The negative ion
methane CI spectra are also characterized by intense
electron capture M⁻ ions, but differ from the
corresponding negative ion thermospray spectra in the
general reduction in the relative intensity of fragment
ions, as well as in the formation of cluster ions $(M+14)^-$
and $(M+16)^-$ (the origins of which we are unable to
determine).

Although the negative ion thermospray fragmentation
is not extensive, it is sufficient to differentiate
between some compounds of identical molecular weight and
similar retention times. Sensitivity is adequate, with a
full spectrum obtainable from as little as 212 ng of
analyte. Due to the intense electron capture M⁻ ion and
the absense of molecular ion adduct ions in these
negative ion thermospray spectra, molecular weight
determination is straightforward.

The similarity of the negative ion thermospray
spectra to the negative ion methane chemical ionization
spectra regarding the facile electron capture process
illustrates the previously reported (5) similarity
between filament or discharge electrode ionization in
thermospray interfaces with classical chemical
ionization. When electron capture is the desired
ionization mode for a conventional negative ion CI
analysis, a non-polar reagent gas (such as methane) is
typically used to avoid other ionization pathways, such
as deprotonation. Solvent composition has an analogous
effect upon the thermospray filament or discharge
electrode ionization electron capture process. For
example, increasing the proton affinity of the
thermospray solvent system by increasing water (with a
gas phase acidity of 1635 kJ/mol (18)) at the expense of
methanol (with a gas phase acidity of 1587 kJ/mol(18))
increases deprotonation ions at the expense of electron
capture ions. At a solvent composition of 50/50 methanol

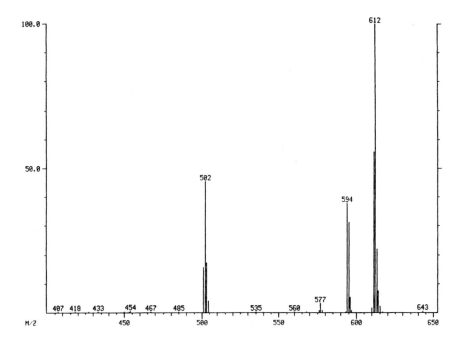

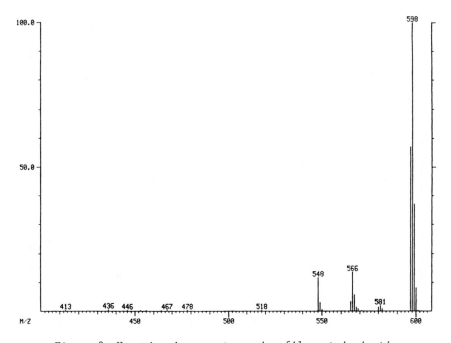

Figure 3. Negative ion spectra using filament ionization.
Continued on next page.

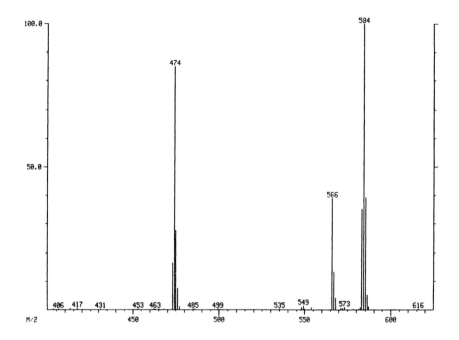

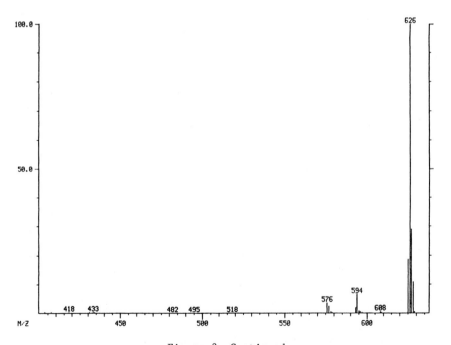

Figure 3. Continued.

Table II. Negative ion spectra[c]

	Alpha	Beta	Gamma	Lambda
M+16⁻	m/z 628(0,1)	m/z 600(0,12)	m/z 614(0,20)	m/z 642(0,52)
M+14⁻	m/z626(0,5)	m/z598(0,10)	m/z 612(0,10)	m/z 640(0,32)
M⁻·	m/z 612(100,100)	m/z 584(100,100)	m/z598 (100,100)	m/z 626(100,100)
M⁻· - H_2O	m/z 594(40,10)	m/z 566(40,10)	m/z 580(3,5)	m/z 608(3,5)
M⁻· - MeOH	not present	not present	m/z 566(15,10)	m/z 594(10,8)
M⁻· -H_2O -MeOH	not present	not present	m/z548(12,7)	m/z 576(7,20)
Retro Diels Alder b	m/z 502(40,5)	m/z 474(80,2)	not present	not present

c. Ions are listed as m/z xxx(A,B) where A is the relative intensity for the ion under filament ionization, thermospray conditions and B is the relative intensity for the ion under methane chemical ionization conditions.

to water, the (M-H)⁻ ion at m/z 611 in a spectrum of
alpha has an intensity of 70% relative to the base peak
due to electron capture at m/z 612. Spectra obtained
under identical instrument conditions from the same
sample using 75/25 methanol/water and 100% methanol
exhibited significantly smaller relative abundaces for
the (M-H)⁻ ion (37% and 25%, respectively). Addition of
as little as 1% acetonitrile or 1% 0.1 M ammonium acetate
to a solvent system of 75% methanol, 25% water shifts the
ionization mechanism from electron capture to
deprotonation. Similar additions with a solvent system
using 90% methanol caused no discernable relative
decrease in electron capture versus deprotonation.
Formation of the strong gas phase Bronsted bases
necessary (19) for deprotonation ionization of the
macrolides is apparently decreased by increasing the
percentage of methanol in the solvent system, thereby
favoring electron capture for the nemadectins alpha,
beta, gamma, and lambda.

Tetracyclines

        Inspection of the total ion chromatogram in Figure 4
from a 15 µL injection of a standard solution of (in
order of elution) 0.0048g/ml 6-demethyltetracycline
(molecular weight 430 amu), 0.0040g/ml tetracycline
(molecular weight 444 amu), 0.0030g/ml declomycin
(molecular weight 464 amu), and 0.0043g/ml
chlortetracycline (molecular weight 478 amu) indicates
that this method provides adequate resolution for
qualitative analysis of tetracyclines. Although similar
quantities of each analyte were injected, a significant
irreversible column retention occurs extent such that
quantitation might not be reliable using this method,
particularly for the chlorinated species. Data obtained
via loop injections (not shown) indicates that each of
these four compounds exhibits approximately the same
sensitivity when the column is bypassed.
    The positive ion thermospray spectra of tetracyclines
are dominated by intense M + H ions (see Figure 5).
Fragmentation for tetracycline and 6-demethyltetracycline
consists of a loss of 18 amu due to dehydation of an
alcohol group and a loss of 44 amu that can be
rationalized as resulting from loss of the dimethylamino
group or possibly from loss of the amide. Declomycin and
chlortetracycline exhibit similar fragmentation losses of
18 amu and 44 amu, with no losses of HCL apparent in the
positive ion spectra. The limit of detection in the
positive ion mode for this method is on the order of 1mg
of analyte, although sensitivity could be enhanced by
scanning a smaller range of masses. Positive ion FAB
spectra are virtually identical to the positive ion
thermospray spectra, with the exception of

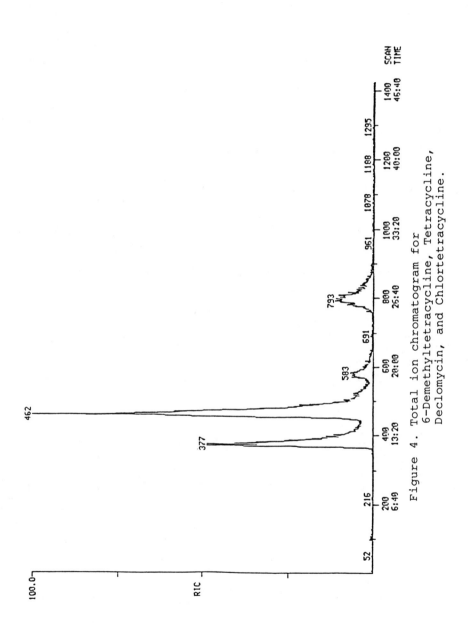

Figure 4. Total ion chromatogram for
6-Demethyltetracycline, Tetracycline,
Declomycin, and Chlortetracycline.

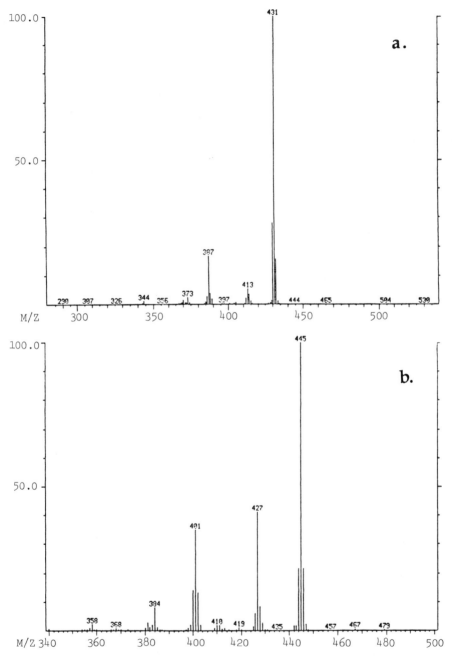

Figure 5. Positive ion spectra of a. 6-Demethyltetracycline and
b. Tetracycline. <u>Continued on next page.</u>

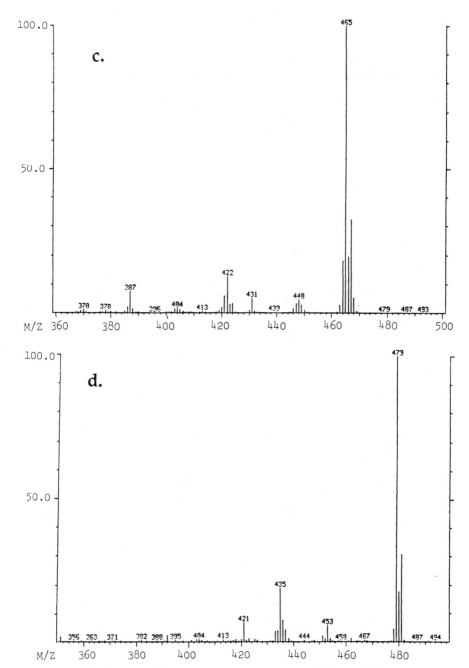

Figure 5. Continued. Positive ion spectra of c. Declomycin and d. Chlortetracycline.

chlortetracycline. Due to matrix effects, an ion at m/z 460 amu of 70% relative intensity to the M+H$^+$ ion corresponding to dehydration of an alcohol occurs. In further contrast to the thermospray spectra, no ion at m/z 435 is observed, indicating that loss of dimethylamine is not facile under these FAB conditions for chlortetracycline.

Total ion current sensitivity in the negative ion mode is approximately an order of magnitude higher than in the positive ion mode. Fragmentation (Figure 6) is considerably more extensive in the negative ion mode than in the positive ion mode. A listing of these fragments is provided in Table III, with probable mechanisms of formation. The high energy, free radical losses of CO and NH$_2$ that occur in the EI spectra ([20]) are absent from the thermospray spectra. The high energy EI conditions are capable of degrading the ring structure, while the mild thermospray conditions leave the ring structure intact. All thermospay fragmentations for these four analytes can be rationalized from chemical ionization-type acid-base decompositions.

The ionization mechanism in the negative ion mode is a mixture of deprotonation and electron capture. The spectra shown in Fig.3 indicate a fairly even distribution of M$\cdot$ and M-1 peaks, although electron capture becomes less important as the source becomes less clean.

CONCLUSION

The optimal thermospray LC/MS analysis of nemadectins uses a methanol/water gradient through a C-18 reversed phase column. External ionization by either filament or discharge electrode provides adequate sensitivity for minor component analysis, providing useful spectra for 200-300 ng of analyte. Positive ion spectra contain diagnostic structural information, due to the abundance of fragment ions produced by dehydration losses and by the particularly informative retro-Diels Alder losses. Negative ion spectra provide unambiguous molecular weight information from the intense M$\cdot$ electron capture ions. Electron capture is enhanced by increasing methanol (relative to water) in the solvent system, while deprotonation is favored by increasing the proportion of water or in some other manner increasing the proton affinity of the solvent relative to the analyte.

The reversed phase (C-4 column) separation using dimethylformamide as the organic modifier with an ammonium acetate buffer provides adequate resolution of tetracyclines for our qualitative analysis application. (It should be noted, however, that a quantitative application of this technique for tetracyclines would not

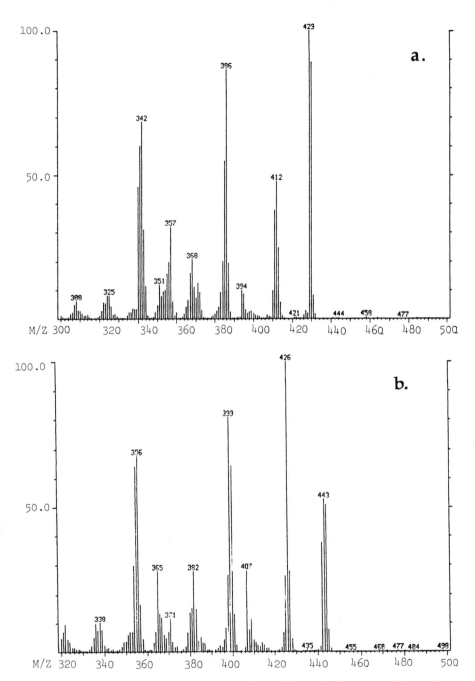

Figure 6. Negative ion spectra of a. 6-Demethyltetracycline and b. Tetracycline. <u>Continued on next page.</u>

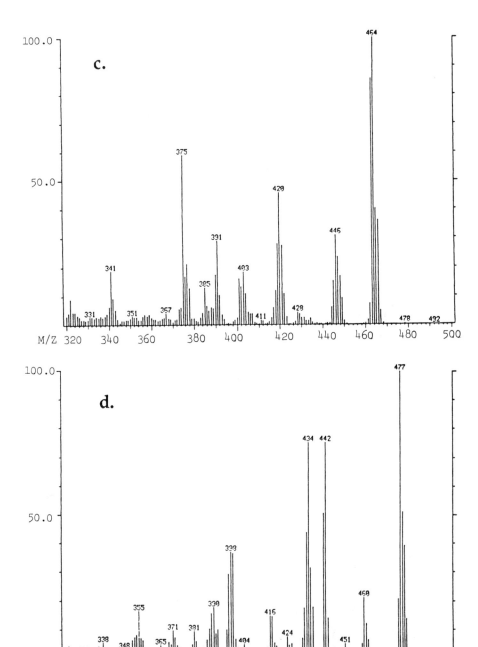

Figure 6. Continued.   Negative ion spectra of c. Declomycin and
d. Chlortetracycline.

Table III. Negative ion Thermospray fragments

	6-demethyltetracycline	tetracycline	declomycin	chlortetracycline
electron capture	430	444	464	478
deprotonation	429	443	463	477
loss of water	412	426	446	460
loss of HCl	–	–	428	442
loss of (CH3)2N	386	399	420	434
loss of water and HCl	–	–	–	424
loss of HCl and (CH3)2N	–	–	–	398
loss of (CH3)2N and CONH2	342	356	375	390
loss of (CH3)2N, water, and CONH2	324	338	–	371

be reliable, due to significant irreversible binding of
some analytes, particularly the chlorinated species, to
the column).the extensive fragmentations of the
tetracyclines in the negative ion thermospray mass
spectra provide useful structural information. The
positive ion mode provides unambiguous molecular weight
information for the tetracyclines due to the intense M+1
ions and lack of extensive fragmentation.

Acknowledgments

    The authors would like to thank Dr. Richard Leese and
Dr. Fred Barbatschi,also of American Cyanamid, for
helpful suggestions.

Literature Cited

1.   Carter, G.T., Nietsche, J.A., Hertz, M.R., Williams,
      D.R., Siegal, M.M., Morton, G.O., James, J.C.,
      Borders,   D.B. J.Antibiotics, 1988, 41 , 519.
2.   Goodman, J.J., Torrey, M.J., Korshalla, J.A., Pinho,
      F.,Testa, R.T. Program and Abstracts of the 27th
      Intersci.Conf.on Antimicrob.Agents Chemother.,1987,
      p. 271.
3.   Blakely, C.M. , Vestal, M.L. Anal. Chem.,
      1984,55,750.
4.   Tway, P.C., Downing, G.V., Slayback, J.R.B., Rahn,
      G.S. and Isense, R.K. Biomed. Mass Spectrom., 1987 ,
      11, 172.
5.   Gartiez, D.A., Vestal, M.L. LC,1985,3, 334.
6.   Tsuji, K.,Robertson, J.H., Beyer, W.F. Anal.Chem.,
      1974, 46, 539.
7.   Tsuji, K.,Robertson, J.H. J.Pharm.Sci., 1975, 110,
      103.
8.   Knox, J.H., Jurand,J. J.Chromatogr., 1975, 110, 103.
9.   Knox, J.H., Pryde, A. J.Chromatogr., 1975, 112, 171.
10.   Chevalier, G., Bollet, C., Rohrbach, P., Risse, C.,
      Caude M., Rosset,R. J.Chromatogr., 1976,124, 754.
11.   Nilsson-Ehle, I., Yoshikawa, J.T., Schotz, M.C.,
      Guze, L.B. Antimicrob.AgentsChemother., 1976, 9,
      754.
12.   White, E.R., Carrol, M.A.,Zarembo,J.E. Bender,
      J.Antibiot., 1975, 28, 205.
13.   De Leenher, A.P., Nelis, H.J.C.F. J.Chromatogr.,
      1977, 140, 293.
14.   De Leenher, A.P., Nelis, H.J.C.F. J.Pharm.Sci.,
      1979, 68, 999.
15.   Mourot, D., Delepine, B., Boisseau, J. and Gayot,G.
      J.Chromatogr., 1980, 190, 486.

16. Leese, R., Barbatschi, F. <u>HPLC Assay For Chlortetracycline in Animal Feed</u>, presented at AOAC National Meeting, Washington, D.C., October, 1983.

17. Voyksner, R.D., Haney, C.A. <u>Anal.Chem.</u>, 1985, <u>57</u>, 991.

18. Kebarle, P. Choudhury, S. <u>Chem. Rev.</u>, 1987, <u>87</u>, 513.

18. Parker, C.E., Smith, R.W., Gaskell, S.J., Bursey, M.M. <u>Anal. Chem.</u>, 1986, <u>58</u>, 1661.

20. Hoffman, D.R. <u>J.Org.Chem.</u>, 1966, <u>31</u>, 792.

RECEIVED October 24, 1989

Chapter 10

Quantification of Endogenous Retinoic Acid in Human Plasma by Liquid Chromatography/Mass Spectrometry

C. A. Huselton, B. E. Fayer, W. A. Garland, and D. J. Liberato

Department of Drug Metabolism, Hoffmann-La Roche Inc., Nutley, NJ 07110

Retinoic acid, an endogenous retinoid, is a potent inducer of cellular differentiation. Because cancer is fundamentally a loss of cellular differentiation, circulating levels of retinoic acid could play an important role in chemoprevention. However, physiological concentrations are typically below the limits of HPLC detection. Sensitive techniques, such as negative chemical ionization (NCI) GC/MS have been employed for quantification, but cause isomerization and also fail to resolve the cis and trans isomers of retinoic acid. Normal phase HPLC can resolve the cis and trans isomers of retinoic acid without isomerization, and mobile phase volatility makes it readily compatible with the mass spectrometer. Based on these considerations, a method combining microbore normal phase HPLC separation with NCI-MS detection was developed to quantify endogenous 13-cis and all-trans retinoic acid in human plasma. The limit of detection was 0.5 ng/ml, injecting only 8 pg of retinoic acid onto the column. The concentration of 13-cis retinoic acid in normal, fasted, human plasma (n=13) was 1.6 +/- 0.40 ng/ml.

Retinoids, vitamin A analogs, support a wide range of physiologic functions. They are necessary for normal vision, growth and reproduction. Retinoids also affect the differentiation and proliferation of both normal and neoplastic cells. They are cytotoxic and have established anti-promoter activity in several model systems ([1-4]). Retinoids are also immunostimulants. Several studies ([5-7]) have indicated that retinoids augment cell mediated cytotoxicity against tumors, increase natural killer cell activity, accelerate graft rejection, increase lymphocyte mitogenesis and augment the cytotoxic and phagocytic activity of macrophages.

0097–6156/90/0420–0166$06.00/0

Because of these activities, retinoids may be important natural chemopreventive agents, and may also have chemotherapeutic value. The majority of human tumors arise in tissues dependent upon retinoids for normal cellular differentiation (8). Retinoids are demonstrated chemopreventive agents in several experimental carcinogenesis models (9-13), including breast, bladder, lung, skin, liver, pancreas, colon and esophagus. Several epidemiological studies (14-16) suggest an inverse relationship between the intake of food with a high vitamin A content and cancer risk. Therefore, an individual's retinoid status may be an important determinant of cancer risk.

Retinoic acid, an endogenous retinoid, is an oxidized metabolite of retinol (vitamin A), and the most potent known inducer of differentiation in vitro (17-19). It is, therefore, most likely the form of vitamin A which promotes normal cellular differentiation. Because cancer is fundamentally a loss of cellular differentiation, physiological concentrations of retinoic acid may play an important role in the etiology of cancer.

However, to assess its role in disease and health, sensitive and specific assays for retinoic acid in biological samples are needed because physiological concentrations are extremely low. Moreover, retinoic acid is sensitive to heat, light and oxygen (20-22). In the presence of these components, it is easily and rapidly isomerized and/or oxidized, thus making quantification difficult.

Most assays for the quantification of endogenous levels of 13-cis and all trans retinoic acid utilize GC/MS. This technique is highly sensitive, but GC isomer resolution is an inherent problem (23). Therefore, unequivocal quantification of cis and trans retinoic acid levels is impossible. The use of HPLC can eliminate isomerization, but lacks sensitivity. Therefore, HPLC in combination with MS should provide a highly sensitive method of quantification without isomerization. In this report, we describe the use of microbore normal phase HPLC/NCI-MS to quantify endogenous levels of 13-cis and all trans retinoic acid in human plasma.

MATERIALS AND METHODS.

MATERIALS. All chemicals and reagents were either reagent or HPLC grade. α-Bromo-2,3,4,5,6-pentafluorotoluene was purchased from Aldrich Chemical Co., Milwaukee, WI. 13-cis and all trans retinoic acid were obtained from Quality Control, Hoffmann-La Roche Inc., Nutley, NJ. [11,12-^{3}H]-all trans retinoic acid, [10,11-^{14}C]-13-cis retinoic acid and 13-cis tetradeuterated retinoic acid were obtained from Dr. A. Liebman, Department of Isotope Synthesis, Hoffmann-La Roche Inc., Nutley, NJ.

HUMAN PLASMA. Human blood was obtained from volunteers who had fasted for eight hours prior to collection. Blood was collected by venipuncture into heparinized Vacutainers. Plasma was prepared by centrifugation at 3000 x g for 20 min at 4°C.

RETINOIC ACID ASSAY. Calibration curves were obtained by adding known amounts of 13-cis (40 μl) and all trans (40 μl) retinoic acid in ethanol to 1 ml of phosphate buffered saline

(PBS). The internal standard, 13-cis tetradeuterated retinoic acid (20 μl) in ethanol, was added to all samples, including the plasma samples. All procedures were performed under yellow lights, including the LC/MS analysis. All glassware was amberized.

EXTRACTION. Retinoic acid was extracted from human plasma or enriched PBS samples after the addition of acidic phosphate (Oldfield, N., Hoffmann-La Roche Inc., personnel communication, 1988). To each 1 ml sample was added 0.5 ml of ethanol and 1 ml of 1M potassium phosphate (pH 3.5), followed by vortexing. After thorough mixing, the samples remained at room temperature for 15 min before extraction with three 1 ml portions of hexane. The layers were separated by centrifugation at 3000 x g for 10 min at 4°C.

DERIVATIZATION. The combined hexane layers were evaporated to dryness under a stream of nitrogen and converted to their penta-fluorobenzyl esters according to the method of Rubio and Garland (24). Derivatized samples were stored in 1 ml of hexane at -20°C.

HPLC. Derivatized samples were evaporated to dryness under a stream of nitrogen and dissolved in mobile phase (50 μl) for analytical reverse phase isolation. Two DuPont Zorbax-ODS (C-18) columns (4.6 mm x 25 cm), in tandem, were connected to a Waters 501 pump. The mobile phase was 100% acetonitrile; the flow rate was 1 ml/min. Samples (50 μl) were injected onto the column using a Waters WISP Model 710B injector. This system was able to resolve 13-cis retinoic acid (t_r = 25 min) from the all trans isomer (t_r = 27 min). The retinoic acid containing fraction (23-29 min) was detected by the UV absorbance of the internal standard at 365 nm and collected. An Applied Biosystems Model 783A variable wavelength detector was used to monitor the UV absorbance, using high sensitivity (AUFS = 0.005).

NORMAL PHASE MICRO-LC/NCI-MS. HPLC purified samples were dried in a Savant Speed-Vac and then dissolved in 50 μl of mobile phase (see below) for LC/MS analysis. A diol column (1 mm x 25 cm, obtained from E.S. Industries, Marlton, NJ) was connected to an Applied Biosystems Microgradient System HPLC pump. Samples (3 μl) were introduced through a Rheodyne Model 8125 injector and eluted isocratically using 15% toluene in hexane at a flow rate of 50 μl/min. Eluate was directly introduced into a modified Finnigan 3200 mass spectrometer equipped with a Teknivent data system. Peak height and ratio data was calculated using a weighted linear analysis program (QSIMPS) (25).

The Finnigan chemical ionization source was modified by the addition of two cartridge heaters. Eluate entered the source through a heated 1/2" probe and excess solvent was removed by a mechanical vacuum pump connected directly opposite the eluate entrance. The interface consisted of 1 meter deactivated silica capillary tubing (ID, 60 μ; OD, 0.008"), led from the outlet of microbore column and threaded through the probe. The probe design has been previously described (26). For all analyses, the probe was operated at 240°C and the source at 250°C. The analyzer pressure was 10^{-5} torr.

RESULTS.

MASS SPECTRA. Figure 1 is the mass spectrum of all trans retinoic acid. The most intense ion is that at m/z 299, the [M-PFB]$^-$ ion. Another ion is seen at m/z 255, representing the loss of PFB and carboxyl groups. The mass spectrum of 13-cis retinoic acid is very similar to that of the all trans isomer with the most intense ion being at m/z 299.

LINEARITY. Since retinoic acid is a normal component in plasma, it is impossible to generate a calibration curve in plasma without first destroying the endogenous retinoic acid. Generation of a plasma blank by UV irradiation was unsatisfactory as it gave erratic results. Therefore, it was decided to obtain a calibration curve by extracting know aliquots of all trans and 13-cis retinoic acid along with internal standard, 13-cis tetradeuterated retinoic acid, from PBS.
    Retinoic acid in PBS was degraded when extracted with methanolic HCL (20,23), so a less harsh extraction procedure using acidic phosphate buffer, pH 3.5, was employed. Figure 2 illustrates the ratio response versus concentration of 13-cis retinoic acid. It is linear throughout the concentration range studied (0.5 to 16 ng/ml). All trans retinoic acid also gave a linear response in the same concentration range. Figure 3 is the selected ion profile of 0.5 ng/ml of both cis and trans retinoic acid and 20 ng/ml tetradeuterated 13-cis retinoic acid extracted from PBS. As can be seen from the selected ion profile, the assay could be extended below 0.5 ng/ml, however, normal physiological values should not fall below this value. The selected ion profile of a blank containing the internal standard only gave a signal from the internal standard.

RECOVERY. A dual label recovery experiment, from plasma and PBS, was performed using ^{3}H-all trans and ^{14}C-cis retinoic acid. The normal extraction procedure was followed up to and including the HPLC purification step. No LC/MS analysis was performed. Aliquots were taken and total radioactivity determined after extraction and derivatization. Fractions (0.5 ml) from the HPLC were collected and counted. Counting was performed using a Beckman Model LC3801 liquid scintillation counter. Radioactivity was corrected for spillover and quench.
    Table I shows the recovery of cis and trans retinoic from plasma and PBS (n=3). Recovery of radioactivity is high throughout extraction and derivatization, but drops off sharply after HPLC

Table I. Recovery of ^{14}C-13-cis and ^{3}H-all trans retinoic acid from PBS or normal human plasma

	% Recovery ± SD (n=3)					
	Extraction		Derivatization		Analytical HPLC	
	cis	trans	cis	trans	cis	trans
PBS	88.6±4.4	69.4±3.8	71.7±3.6	57.6±3.0	32.4±4.4	26.7±3.6
Plasma	95.5±1.7	78.8±1.9	78.2±3.9	65.1±4.4	29.0±3.3	19.5±2.9

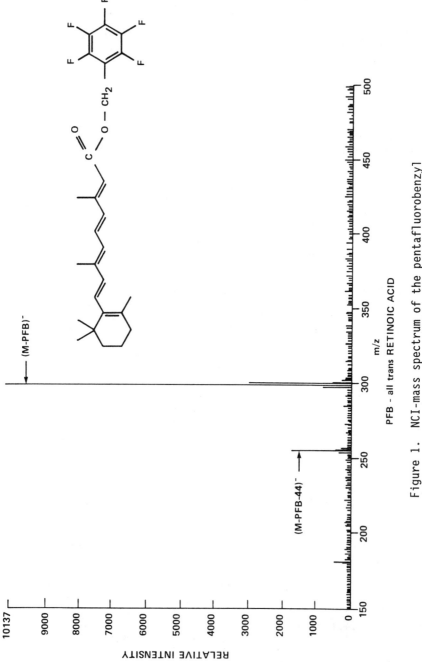

Figure 1.   NCI-mass spectrum of the pentafluorobenzyl derivative of all trans retinoic acid.

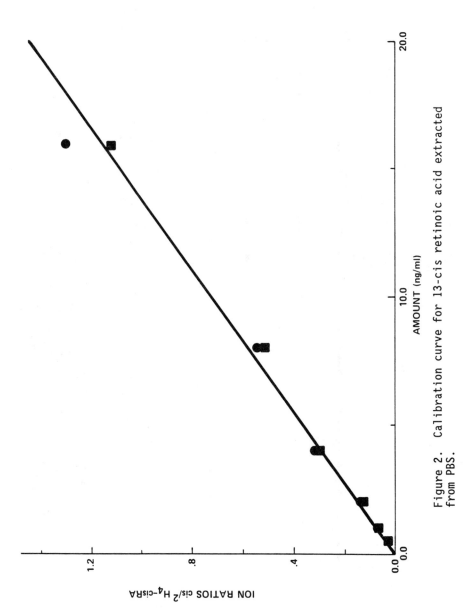

Figure 2. Calibration curve for 13-cis retinoic acid extracted from PBS.

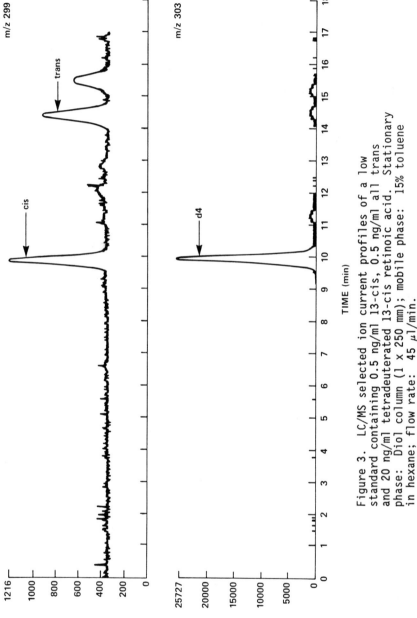

Figure 3. LC/MS selected ion current profiles of a low standard containing 0.5 ng/ml 13-cis, 0.5 ng/ml all trans and 20 ng/ml tetradeuterated 13-cis retinoic acid. Stationary phase: Diol column (1 x 250 mm); mobile phase: 15% toluene in hexane; flow rate: 45 µl/min.

purification. It should be noted that the radioactivity calculated after extraction and derivatization represents the entire radioactivity in the sample, whereas, the radioactivity after HPLC purification represents only that fraction taken on to LC/MS analysis.

There appears to be a differential recovery of cis over trans retinoic acid throughout all the steps of the extraction procedure. However, it is more pronounced after HPLC purification, especially in the plasma samples. In these samples the recovery of all trans retinoic acid was less than that from the PBS samples. This observation most likely leads to a slight underestimation of endogenous all trans retinoic acid in plasma, when using the 13-cis retinoic acid internal standard for quantification.

Isomer interconversion due to assay manipulations could also be determined from the recovery experiment. Isomerization data was obtained from the radioactivity profile of the HPLC purification step. The amount of cis isomerizing to trans was 3%. The same amount of trans retinoic acid isomerized to cis. The assay causes a small amount of isomerization but to an equal extent for both isomers.

ACCURACY. Assay accuracy was examined by performing a standard addition experiment. To plasma was added either 0, 5 or 10 ng/ml of 13-cis or all trans retinoic acid, followed by extraction (n=10). A mixing experiment (n=4), in which plasma was fortified with both cis and trans retinoic acid, was also performed. No mixing effect was observed. The internal standard was 20 ng/ml throughout. Table II shows the results of this experiment. The observed values correlate quite well for 13-cis retinoic acid. However, the observed values for all trans retinoic acid were less than those calculated. This is most likely due to the somewhat poorer recovery of the all trans isomer from plasma.

Table II. Standard addition of 13-cis or all trans retinoic acid to normal human plasma

		+ 5 ng/ml		+ 10 ng/ml	
	Background	Obs.	% Error	Obs.	% Error
cis	1.6 ± 0.1 ng/ml	7.0 ± 0.2 ng/ml	+6.1	11.8 ± 0.4 ng/ml	+1.7
trans	1.4 ± 0.1 ng/ml	5.1 ± 0.5 ng/ml	-20.3	8.0 ± 0.4 ng/ml	-29.8

ENDOGENOUS LEVELS IN HUMAN PLASMA. Figure 4 is the selected ion current profile of endogenous retinoic acid extracted from the plasma of a male volunteer. No differences were noted in the profiles between male and female volunteers. All profiles were quite similar in appearance. Table III lists the individual concentrations of endogenous plasma cis and trans retinoic acid of

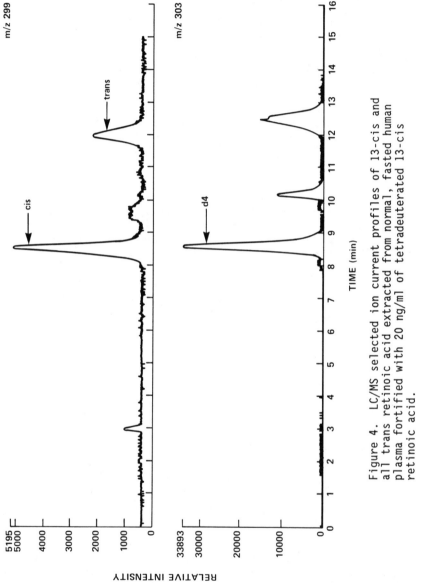

Figure 4. LC/MS selected ion current profiles of 13-cis and all trans retinoic acid extracted from normal, fasted human plasma fortified with 20 ng/ml of tetradeuterated 13-cis retinoic acid.

the thirteen volunteers. The concentration range of either cis or trans retinoic acid is very small. The mean (± SD) values were 1.6 ± 0.4 and 0.9 ± 0.2 ng/ml for 13-cis and all trans retinoic acid, respectively.

Table III. Individual concentrations of 13-cis and all trans retinoic acid in fasted, normal human plasma

13-CIS RETINOIC ACID

Subject Number	Amount (ng/ml)	Subject Number	Concentration (ng/ml)
F1	2.2	M1	1.5
F2	1.9	M2	1.5
F3	0.90	M3	1.5
F4	1.5	M4	1.2
F4	1.0	M5	1.5
F6	1.7	M6	2.2
		M7	1.8

Mean Concentration ± SD   1.6 ± 0.4

ALL TRANS RETINOIC ACID

Subject Number	Amount (ng/ml)	Subject Number	Concentration (ng/ml)
F1	0.8	M1	0.9
F2	1.1	M2	0.9
F3	0.9	M3	0.8
F4	1.0	M4	0.7
F4	0.7	M5	1.2
F6	1.2	M6	1.0
		M7	1.0

Mean Concentration ± SD   0.9 ± 0.2

F: female; M: male.

DISCUSSION.

Retinoic acid is involved in only two of the known functions of Vitamin A, namely growth and cellular differentiation. Retinoic acid is a potent inducer of cellular differentiation, in vitro (17-19), and may be necessary for normal gene expression (17,27). Several nuclear transcription factors for retinoic acid have been identified (28,29) and the associated genes products may be necessary to prevent cellular transformation. Therefore, retinoic acid may be necessary for homeostasis. However, because of its potency, it exists in low concentrations in biological systems.

Therapeutic doses of retinoic acid can be toxic and are teratogenic (30-31). It can cross the placenta to the fetus (31,32), resulting in fetal deformities. Barua and Olson (33) have suggested that the biologically inactive form of retinoic acid is the ß-glucuronide. The glucuronide is less toxic and not teratogenic (Olson, A.L., Iowa State University, 1989). When administered to pregnant rats, it was found not to cross the placenta to the fetus (Olson, A.L., Iowa State University, 1989). The ß-glucuronide may be converted to the active form, retinoic acid, by a glucuronidase (33,34).

Several investigators have developed assays for the quantification of all trans retinoic acid in human blood, plasma or serum. Nelson et al. (35) developed a colorimetric assay, but could not detect retinoic acid under physiological conditions. DeRuyter et al. (36) found 1 to 3 ng/ml of all trans retinoic acid in fasted human serum using normal phase HPLC. Chiang (37) validated an assay for all trans retinoic acid in plasma, but could not detect any under normal physiological conditions using 10 ml of plasma. The limit of detection was 1 ng/ml. Another HPLC assay was developed by DeLeenheer et al. (38), who found serum concentrations of all trans retinoic acid to be 3.5 ng/ml, using 3.5 ml of serum. The limit of sensitivity was 1 ng/ml.

Mass spectrometric analysis using negative chemical ionization techniques affords greater sensitivity over electron impact methods in that a very intense ion expressing the molecular weight of the compound of interest is produced. Quantification using the stable isotope dilution method is exact and reproducible. The method does not rely on the efficiency of extraction, but rather on the ratio between the compound of interest and the internal standard, which remains constant in an individual sample.

Napoli et al. (23) developed a sensitive assay based on negative chemical ionization mass spectrometry to quantify retinoic acid in human plasma. Endogenous levels of all trans retinoic acid in plasma were 4.9 ng/ml, using a 0.1 ml sample. The limit of detection was less than 1 ng/ml. Direct quantification of 13-cis retinoic acid was impossible due to the inability of the GC to resolve the isomers. Barua and Olson (33) described a method to quantify all trans retinoic acid in serum using reverse phase HPLC. They detected 1.8 ng/ml of the all trans isomer, using a 2 ml serum sample and a non-acidic extraction procedure.

We have developed a very sensitive assay which can quantify both 13-cis and all trans retinoic acid in the same plasma sample. Only 1 ml of plasma is necessary for analysis, with a limit of quantification of 0.5 ng/ml. The assay is linear for both cis and trans retinoic acid, and there is virtually no interconversion of the two isomers by assay manipulations. However, the assay does slightly underestimate the amount of all trans retinoic acid present due to the differential recovery of this isomer from plasma as opposed to recovery from PBS. This will be corrected in future work by the addition of a stable isotope labelled all trans retinoic acid internal standard for quantification.

In our studies, plasma concentrations of all trans retinoic acid ranged from 0.7 to 1.2 ng/ml (mean of 0.9 ng/ml); the 13-cis concentrations ranged from 0.9 to 2.2 ng/ml (mean of 1.6 ng/ml).

The concentration of all trans retinoic acid in plasma as determined from our assay correlates well with other studies, using human serum (36-38), but is lower than that found by Napoli et al. (23).

The higher concentration reported by Napoli et al. (23) may be due to hydrolysis of the glucuronide metabolite since retinoyl glucuronide is hydrolyzed by acid or base to retinoic acid (33,34). The extraction procedure of Napoli et al. (23) utilized 2N HCl which could potentially convert most of the retinoyl glucuronide to retinoic acid. Therefore, the concentration of all trans retinoic acid as reported by this assay could represent the sum of these two retinoids, and not that of the free acid (33). Our assay uses less harsh acidic conditions for extraction which should prevent hydrolysis of the glucuronide, resulting in a lower plasma concentration of the all trans isomer as reported by Napoli et al. (23).

Future work includes defining endogenous plasma levels of 13-cis and all trans retinoic acid in an expanded population and determining whether normal plasma levels are affected by disease or nutritional state. Finally, the assay will be expanded to measure the major metabolites of retinoic acid (4-hydroxy, 4-oxo, glucuronide conjugates, etc.).

LITERATURE CITED

1. Bollag, W. Experimentia 1971, 27, 90-92.
2. Borek, C.; Smith, J.E. J. Cell Biol. 1978, 79, CU341.
3. Lotan, R., J. Cell Biol. 1978, 79, CD144.
4. Mufson, R.A.; DeFeo, D.; Weinstein, I.B. Mol. Pharmacol. 1978, 16, 569-578.
5. Watson, R.R. In Vitamins and Cancer-Human Cancer Preventions by Vitamins and Micronutrients; Humana Press, Inc., 1985; pp. 439-51.
6. Watson, R.R.; Rybski, J. In Nutrition and Immunology; Chandra, R.K., Ed.; A.R. Liss, NY, 1988; pp. 89-99.
7. Moriguchi, S.; Werner, L.; Watson, R.R. Immunology 1985, 56, 169-177.
8. Moore, T. In Vitamin A; Elsevier Publishing Corp., Amsterdam, 1957; pp. 295-338.
9. Boutwell, R.K. Cancer Res. 1983, 43, 2465s-2468s.
10. Moon, R.C. and Itri, L. In The Retinoids; Sporn, M.B.; Roberts, A.B.; Goodman, D.S., Eds.; Academic Press, NY, 1984; Vol. 2, pp. 327-371.
11. Daoud, A.H.; Griffin, A.C. Cancer Lettr. 1980, 9, 299-304.
12. Longnecker, D.S.; Curphey, T.J.; Kuhlman, E.T.; Roebuck, B.D. Cancer Res. 1982, 42, 19-24.
13. Sporn, M.B.; Squire, R.A.; Brown, C.C.; Smith, J.M.; Wenk, M.L.; Springer, S. Science 1977, 195, 487-489.
14. Mettlin, C.; Graham, S.; Swanson, M. JNCI 1979, 62, 1435-1438.
15. Mettlin, C.; Graham, S. Am. J. Epidemiol. 1979, 110, 255-263.
16. Hirayama, T. Nutr. Cancer 1979, 1, 67-81.
17. Lotan, R. Biochim. Biophys. Acta 1980, 605. 33-91.
18. Strickland, S.; Sawey, M.J. Dev. Biol. 1980, 78, 76-85.

19. Williams, J.B.; Shields, C.O.; Brettel, L.M.; Napoli, J.L.
    Anal. Biochem. 1987, 160, 267-274.
20. Napoli, J.L. In Methods in Enzymology, Colowick, S.; Kaplan,
    N.O., Eds; Academic Press, NY, 1986; Vol. 123, pp. 112-124.
21. Vecchi, M.; Vesely, J.; Oesterhelt, G. J. Chromatogr. 1973,
    447.
22. Tayler, R.F.; Davies, B.H. J. Chromatogr. 1975, 103, 327.
23. Napoli, J.L.; Pramanik, B.C. Williams, J.B.; Dawson, M.I.;
    Hobbs, P.D. J. Lipid Res. 1985, 26, 387-392.
24. Rubio, F.; Garland, W.A. J. Chromatogr. 1985, 339, 313-320.
25. Garland, W.A.; Barbalas, M.P.; Hess, J.R. Trends Anal. Chem.
    1986, 5, 132-138.
26. Rubio, F.R.; Fukuda, E.K.; Garland, W.A. Drug Metabol. Dispos.
    1988, 16, 773-777.
27. Roberts, A.B.; Sporn, M.B. in The Retinoids; Sporn, M.B.;
    Roberts, A.B., Eds.; Academic Press, NY, 1988; pp. 236-238.
28. Petkovich, M.; Brand, N.J.; Krust, A.; Chambon, P. Nature
    1987, 330, 444-450.
29. Brand, N.; Petkovich, M.; Krust, A.; Chambon, P.; deThe, H.;
    Marchio, A.; Tiollais, P.; Dejean, A. Nature 1988 332, 850-853.
30. Turton, J.A., Hicks, R.M.; Gwyne, J.; Hunt, R.; Hawkey, C.M.
    In Retinoids, Differentiation and Disease; Ciba Foundation
    Symposium, Pitman, London, 1985; pp. 225-251.
31. Gelen, J.A.G. Crit. Rev. Toxic. 1979, 6, 351-375.
32. Kistler, A. Teratology 1981, 23, 25-31.
33. Barua, A.B.; Olson, J.A. J. Clin. Nutri. 1986, 43, 481-485.
34. Barua, A.B.; Batres, R.O.; Olson, J.A. Biochem. J. 1988, 252,
    415-485.
35. Nelson, E.C.; Dehority, B.A.; Teague, M.S. Anal. Biochem. 1965,
    11, 418.
36. DeRuyter, M.G.; Lambert, W.E.; DeLeenheer, A.P. Anal. Biochem.
    1979, 98, 402-409.
37. Chiang, T-C. J. Chromatogr. 1980, 182, 335-350.
38. DeLeenheer, A.P.; Lambert, W.E.; Claeys, I. J. Lipid Res. 1985,
    23, 1362-1367.

RECEIVED October 6, 1989

Chapter 11

Liquid Chromatography/Mass Spectrometry in Bioanalysis

W. M. A. Niessen[1], U. R. Tjaden[1], and J. van der Greef [1,2]

[1]Division of Analytical Chemistry, Center for Bio-Pharmaceutical Sciences, P.O. Box 9502, 2300 RA Leiden, Netherlands
[2]TNO–CIVO Institutes, P.O. Box 360, 3700 AJ Zeist, Netherlands

Combined liquid chromatography/mass spectrometry (LC/MS) is a powerful tool in both qualitative and quantitative bioanalysis. This paper deals with three aspects of bio-analysis with LC/MS. First, general optimization strategies of thermospray LC/MS are discussed. Next, two examples of bioanalytical applications of LC/MS are given. The first example deals with the confirmation of the presence of aminonitrazepam in a post-mortem blood sample by means of thermospray LC/MS and LC/MS/MS. The second example deals with the analysis of metoprolol enantiomers after a chiral separation using a special application of coupled column chromatography, i.e. the phase-system switching approach, in combination with moving belt LC/MS.

Combined liquid chromatography/mass spectrometry (LC/MS)can play an important role in both qualitative and quantitative bioanalysis. LC/MS can be performed with a number of interfaces. Three interfaces are presently available in our laboratories: i.e., the thermospray interface (TSP), the moving-belt interface (MBI), and continuous-flow fast atom bombardment (CF-FAB). These interfaces are supple-mentary with respect to their applicability and the type of information that can be obtained.

In this paper some important aspects of LC/MS in bioanalysis are discussed. In order to achieve best performance in thermospray LC/MS optimization of the various experimental parameters is necessary. As a result of systematic studies on the influence of the various parameters an optimization strategy is developed, which will be discussed. Two recent bioanalytical applications are discussed. The applications are concerned with confirmation and identification of compounds in biological matrices (qualitative aspects) and with target-compound analysis. The examples have been selected in order

0097–6156/90/0420–0179$06.00/0
© 1990 American Chemical Society

to demonstrate some of the general approaches used in our laboratory
to solve (bioanalytical) problems with LC/MS.

Experimental

Equipment and Solvents. The LC/MS system applied in the aminonitrazepam project consisted of a Model 2150 LC pump (LKB, Bromma,
Sweden) and a Finnigan MAT TSQ-70 tandem mass spectrometer (San
José, CA, USA) equipped with a Finnigan MAT thermospray interface.
The experiments were performed in flow-injection (column by-pass)
mode (FIA). The same system was also used in the development of
optimization strategies for thermospray LC/MS. A mixture of 20%
(v/v) methanol in 50 mmol/l ammonium acetate in water (1.5 ml/min)
was used as the solvent in thermospray buffer ionization, while
in discharge-on experiments 80% methanol in water was used. The
discharge voltage was 1 kV. In MS/MS experiments air was used as
a collision gas; the collision energy was optimized.
     The separation of the metoprolol enantiomers was performed on
a $_1$-acid-glycoprotein column (100 mm x 4 mm ID) (ChromTech AB,
Stockholm, Sweden) with a mobile phase of 0.25% 2-propanol in
20 mmol/l phosphate buffer (pH=7) at a flow-rate of 0.8 ml/min.
The trapping columns were Guard-Pak^{tm} C18 cartridges (4 mm x 6 mm ID)
(Waters Assoc., Milford, MA, USA) (1). Desorption of the enantiomers
from the trapping column was performed with 0.4 ml/min of methanol,
which stream was directed to a Finnigan MAT HSQ-30 hybrid mass
spectrometer equipped with a moving belt interface. Operating parameters of the moving belt interface were a solvent evaporator temperature of 175°C, a belt speed of 3 cm/s, a sample evaporator temperature setting of 7, and the clean-up heater 50%. Mass spectrometry
was performed after ammonia chemical ionization.

Sample Pretreatment. Aminonitrazepam was extracted from the post-
mortem blood by means of a solid-phase isolation on an Extrelut
cartridge column (Merck, Darmstadt, FRG). The cartridge was pre-
treated before use. It was washed with a 1:1 methanol dichloromethane mixture, dried and treated with ammonia vapor. The post-
mortem blood sample (1 ml) was diluted with 0.5 ml water and sucked
through the cartridge. The column was eluted with 8 ml of diethyl
ether. The eluate was evaporated to dryness and redissolved in
methanol.
     The solid-phase isolation used in the sample pretreatment of
metoprolol from plasma was described in detail elsewhere (2),

Optimization Strategies in Thermospray

The signal in thermospray LC/MS is the result of an interweaving
play of many interdependent experimental parameters. Systematic
studies have been undertaken in order to optimize the sensitivity
and the reproducibility of thermospray LC/MS for bioanalytical
applications (3). During and as a result of these systematic
studies optimization strategies for thermospray LC/MS have been
developed.

First, some general guidelines can be given. A clean ion source
and repeller electrode are obligatory for the optimum performance of
the system. For thermospray buffer ionization the concentration of
ammonium acetate (or any other volatile buffer with good gas-phase
proton transfer properties) in the eluent should be higher than the
analyte concentration; with 10-100 mmol/l ammonium acetate in the
mobile phase this condition is nearly always fulfilled. When the
addition of ammonium acetate to the LC eluent is undesirable, either
other ionization modes (filament-on or discharge-on) should be
selected, or post-column addition of ammonium acetate should be
considered. The mobile phase should not contain modifier contents
over 40%, unless another ionization mode, e.g. discharge-on, is used.
   The optimization strategy can be summarized as follows (3).
Before starting the experiments for a particular application the
performance of the thermospray has to be checked; eventually a good
spray can be obtained by squeezing the end of the capillary at
ambient temperature and pressure. Under vacuum conditions the vaporizer temperature has to be adjusted to the flow-rate and the mobile
phase composition, which are in most cases determind by the chromatographic system. Studying the influence of the repeller potential
on the solvent background spectrum can be used to ascertain good
performance of the thermospray interface. When no response of the
total ion current on the repeller potential variation is observed,
either the spray performance is not well adjusted, or the ion source
and the repeller electrode are seriously contaminated. Cleaning of
the ion source is the only appropiate solution to contamination.
   When proper performance of the spray and the repeller electrode
is established, fine tuning of the vaporizer temperature for the
analyte(s) of interest can be performed. Within a temperature range
of ca. 30 °C, depending on the flow rate and the mobile phase
composition and to some extent to the age and history of the vaporizer, the vaporizer temperature has to be optimized, which sometimes
results in a tenfold gain in analyte signal. Especially for thermo-
labile compounds the vaporizer temperature is an important parameter;
adjustment of the vaporizer temperature can have a considerable
influence on the mass spectrum of the analyte, because for thermo-
labile compounds mixed spectra of analyte and its decomposition
products are obtained. When set at a potential between 0 and 100 V
the repeller electrode in most cases has not much influence on the
analyte signal. Therefore the repeller potential only needs systematic optimization when minor gains in sensitivity are needed.
   These general guidelines can be applied in both qualitative and
quantitative analysis. Optimization approaches are generally of more
value in quantitative than in qualitative analysis. In order to obtain the lowest possible detection limits in thermospray LC/MS the
system has to be optimized for each compound or group of compounds.
This optimization procedure has to be repeated before each experimental series, because the parameters which cannot be systematically
adjusted, such as the condition of the vaporizer and the degree of
contamination, can dramatically influence the optimum settings of
the other parameters. The use of isotopically labelled internal
standards is advised in order to avoid the influence of the un-
predictable parameters. The usefulness of extensive optimization in
qualitative analysis with thermospray LC/MS can be questioned.

Sometimes optimization on one of the sample constituents can be useful, while in other cases the various components in the mixture show
widely different thermospray behavior. In many applications standards
are simply not available; the FIA injection of the sample can be
useful, although in many cases the analysis must be focused on the
minor components in the mixture, which will not be clearly detected
in FIA mass spectra.

The selection of an appropriate mobile phase is another aspect
of concern. Most often the commonly used LC mobile phase is not
compatible with thermospray LC/MS, for instance owing to the use
of non-volatile buffers can be left out or replaced by volatile ones.
In other cases the buffers are present for retention time reproducibility, which mostly is not very important for identification. In
other applications however a correspondence between UV or fluor-
escence peaks and MS identification is obligatory, which makes
mobile phase changes unattractive. In this respect it is often over-
looked that LC-UV and LC/MS give different responses as a result of
different detection principles. For not too complex samples a UV
photo-diode array detector can be used to link up the chromatographic
peaks obtained under different mobile phase conditions. To cut short,
despite many successes also many potential problems are met in LC/MS
to which tailor-made creative solutions are needed.

Aminonitrazepam

A man was killed by drowning, probably after being intoxicated. From
judicial inquiries the intoxicating agent is expected to be nitrazepam. Aminonitrazepam (MW 251) is the major metabolite of nitrazepam
(MW 281). In the GC/MS analysis of the postmortem blood-extract with
selective ion monitoring a peak due to a compound with a molecular
weight of 251 was detected, but no mass spectrum could be obtained.
Nitrazepam itself was not detected. Therefore, the sample was
brought to our laboratory to be analyzed by other methods. TSP/MS
and TSP/MS/MS in buffer ionization mode was applied in solving this
problem.

The TSP/MS performance, the repeller potential and the vaporizer
temperature have been optimized with standard solutions of nitrazepam and aminonitrazepam in the usual way described above. In the TSP
spectrum of aminonitrazepam three peaks appear: the protonated
molecule at m/z=252, a fragment peak at m/z=222, which is due to the
loss of formaldehyde, and a fragment peak at m/z=213, which is most
probably a thermally induced hydrolysis product. About 100 ng per
injection (20 µl) was necessary to obtain a spectrum with a
satisfactory signal-to-noise ratio.

Since a better confirmation of the presence of aminonitrazepam
in the blood-extract is achieved when more structural information
is available, the possibilities of TSP/MS/MS have been investigated.
The use of a specific and selective collisionally induced fragmentation reaction would in principle be a powerful probe. Furthermore,
better detection limits can be obtained in MS/MS for compounds
present in complex matrices.

The daughter spectrum of the protonated aminonitrazepam given in Figure 1, contains several peaks. The intense fragment peak at m/z=121 is selected for use in selective reaction monitoring (SRM) in the analysis of the post-mortem sample. The reconstructed chromatograms (Figure 2) obtained with SRM (m/z=252 --> m/z=121) after subsequently injecting 5 µl of the blood-extract and 5µl of blank eluent clearly shows a signal due to the sample and no signal from the blank. In this way the presence of the aminonitrazepam in the blood-extract was sufficiently confirmed.

Two questions of more or less academic interest remained: What is the identity of the peak at m/z=213, appearing in the TSP/MS spectrum of aminonitrazepam, and of the peak at m/z=121, appearing in the daughter spectrum of aminonitrazepam and used in SRM in analyzing the post mortem sample.

The peak at m/z=213 corresponds theoretically to a loss of 39 amu from the protonated aminonitrazepam, which cannot be explained by a normal fragmentation reaction. The peak at m/z=213 increases in itensity with increasing vaporizer temperature, indicating thermal activation. A possible explanation is that after the loss of formaldehyde a nitrile compound is formed, which hydrolyzes resulting in a apparent loss of 9 amu (-C≡N --> -OH). A similar hydrolysis reaction has been observed with several other nitriles, for instance phenylalanine-nitrile with a molecular weight of 160 gives only a small protonated molecule at m/z=161, but an intense fragment at m/z=152. The daughter spectrum of m/z=213 gives no clear answer: it shows peaks at m/z=105, which might be due to $C_6H_5-N=CH_2+$, and m/z=77, which is due to the phenyl group. Further investigations using other benzodiazepines are necessary to prove this process.

In the elucidation of the structure of the daughter ion at m/z=121 it appeared possible to make use of repeller-induced dissociation in the discharge-on mode (4-6) in combination with tandem mass spectrometry. In the discharge-on mode using for instance methanol in water as a solvent often useful structural information can be obtained by applying a potential of 100-200 V at the repeller electrode, which is placed opposite to the sampling cone. At low repeller potentials in the reagent gas from a methanol-water solvent micture protonated methanol (clusters) are the most abundant ions, while at higher repeller potentials protonated water becomes most abundant. The latter has a lower proton affinity and thus will result in a somewhat harder ionization, i.e. more fragmentation. Besides these changes in the reagent gas spectrum the fragmentation at higher repeller potentials might also be induced by collisions within the high pressure ion source (4-6).

The spectrum of aminonitrazepam at low repeller potential is similar to that observed in thermospray buffer ionization. At high repeller potential the spectrum, which is shown in Figure 3, largely resembles the CID daughter spectrum (Figure 1). A strong peak at m/z=121 is observed, together with some minor other fragments (Figure 3a). The daugther spectrum of this peak at m/z=121 is given in Figure 3b, showing peaks at m/z=104 and 94, which can be explained as protonated benzonitrile and protonated aniline, respectively, leading to the structure given in Figure 3b.

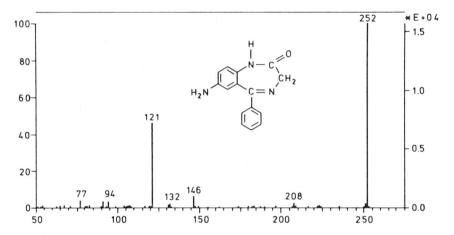

Figure 1. Daughter spectrum of aminonitrazepam. Direct injection in thermospray MS, solvent 20% methanol in 50 mmol/l ammonium acetate, repeller 50V, vaporizer 120°C, collision gas: air at 0.05 Pa, collision energy 60 eV.

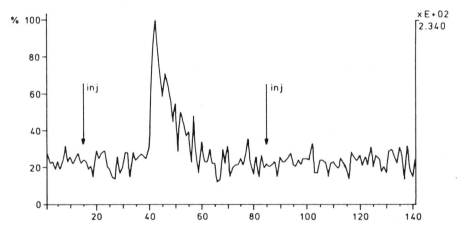

Figure 2. Reconstructed chromatogram from SRM (m/z=252--> m/z=121) after successive injection of post-mortem blood extract (first injection) and blank sample (second injection). Conditions: see figure 1.

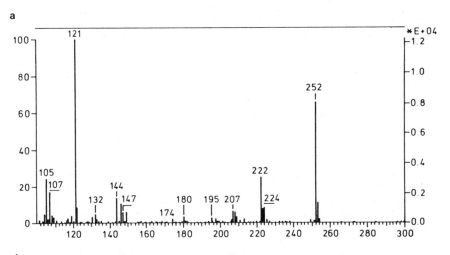

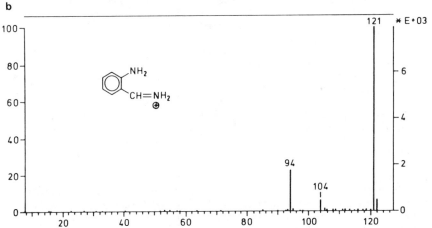

Figure 3.
(a) Spectrum of aminonitrazepam in discharge-on mode at
     repeller potential of 100 V. Conditions: solvent 80%
     methanol in water, vaporizer 100°C, discharge 1 kV.
(b) MS/MS daughter spectrum of the m/z=121 peak from aminoni-
     trazepam and structure proposal. Conditions: collision
     energy 20 eV. other conditions: see (a).

Phase-system switching

One of the major problems in LC/MS interfacing is the fact that most
of the present LC/MS interfaces do not allow the use of mobile phase
additives, such as buffers, ion-pairing, and complexing agents,
while these are frequently used in the daily practice of bioanalysis
with liquid chromatography. Online post-column extraction devices
have been described to solve this problem (7, 8), but they are not
widely applied for this purpose. Recently, another approach has been
demonstrated which is based on valve switching techniques: the so-
called phase-system switching (PSS) (9, 10). A schematic diagram of
the PSS approach is given in Figure 4. In PSS the peak of interest
is heart-cutted from the chromatogram developed on the analytical
column with a mobile phase containing for instance a phosphate buffer
and adsorbed on a short column packed with a material that exhibits
larger capacity ratios for the compound of interest in the mobile
phase used in the analytical separation than the material of the
analytical column. This trapping column is then washed to remove the
non-volatile buffer components, and in some applications also dried,
to remove the excess of washing water. In the last step of the procedure the compound of interest is desorbed from the trapping
column with a solvent that is favorable for the LC/MS interface and
with an appropriate flow-rate.
     As a demonstration of the PSS approach some preliminary results
will be shown for the analysis of the enantiomers of the betablocker
metoprolol (protonated molecule at m/z=268). The separation of the
enantiomers is achieved by means of a chiral column and a phosphate
buffer. The enantiomers are subsequently adsorbed on two C18 trapping
columns, which are washed, dried and successively desorbed with
methanol, which is directed to a moving belt interface and analyzed
under CI conditions. In these studies an isotopically labelled internal standard ($^2H_6$-metoprolol, protonated molecule at m/z=274) is
used. The resulting peaks are shown in Figure 5. These figures
immediately show some problems encountered in this study. The moving
belt shows very high memory effects, a problem which is sometimes
observed. Deactivation of the belt in this case gave no improvement.
As a result only about 50% of the sample is contained in the first
peak. High blank signals are observed, which can be attributed to
the fact that on desorption of the trapping columns the solvent flow
to the moving belt is started; the moving belt runs dry during the
previous steps. MS/MS has been demonstrated to be a good solution
for these types of base-line disturbance (2).
     Nevertheless, Figure 5 demonstrates that the enantiomeric
separation using a phosphate buffer in the mobile phase can be
coupled via the PSS approach on-line to an LC/MS moving belt interface. Other examples of the PSS approach or similar procedures with
other compounds and other LC/MS interfaces have been described
(2, 9-14). Besides the actual phase-system switching, which enables
the choice of the most favorable solvent for a particular interface,
the PSS approach offers some other features as well. The desorption
flow-rate used can be adjusted to the capabilities of the LC/MS
interface applied. While in the present example with the moving belt
a flow-rate of 0.4 ml/min of methanol was used, desorption has also
been demonstrated with 1.2 ml/min for a thermospray interface(2),

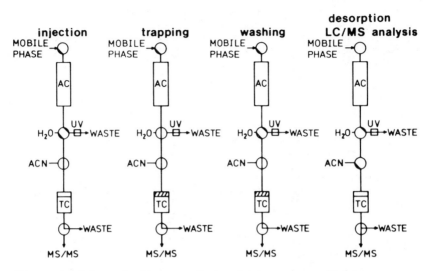

Figure 4. Schematic diagram of the phase-system switching approach applied in the moving belt LC/MS analysis of metropolol enantiomers. AC=analytical column, UV=UV-detector, ACN=desorbing eluent (see text), and TC=trapping column.

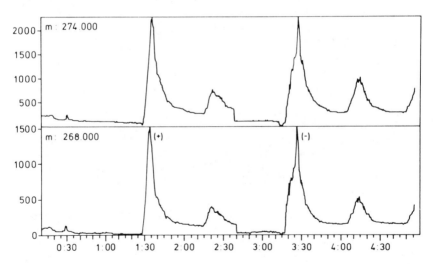

Figure 5. Peaks of the (+)- and (-)-metoprolol enantiomers after chiral separation, phase-system switching and moving belt LC/MS. Conditions: see text.

and with 7 μl/min for a continuous-flow fast-atom bombardment system
(4). Another feature of the PSS approach is the peak compression,
i.e. the broad chromatographic peak from the analytical column is
compressed on the trapping column. Desorption in a strong eluent
results in a narrow band. The peak compression effects have been
studied in more detail recently; the results will be reported elsewhere (13).

Conclusions

The two examples discussed show the bioanalytical applicability of
LC/MS, both in qualitative analysis and in target compound analysis.
Although the moving belt interface owing to its EI capabilities is
ideally suited for the identification of unknown compounds and is
used often for that purpose, identification problems can sometimes
also be solved with thermospray LC/MS, for instance by applying
repeller-induced fragmentation, and LC/MS/MS. The PSS approach is
an example that indicates that LC/MS still can be improved. A more
elaborate discussion on the so-called multidimensional approaches
in LC/MS is given elsewhere (14).

Acknowledgments

The aminonitrazepam project was performed in corporation with
dr. A.M.A. Verweij from the Gerechtelijk Laboratorium (Laboratory of
Forensic Sciences) in Rijswijk, The Netherlands. The metoprolol
project was performed in corporation with A. Walhagen from Technical
Analytical Chemistry at the University of Lund, Sweden and dr. L.-E.
Edholm from the Bioanalytical Laboratory of AB Draco (Subsidiary of
AB Astra) in Lund, Sweden. From within our laboratories C.E.M.
Heeremans, R.A.M. van der Hoeven and E.R. Verheij are thanked for
their contributions to the projects discussed.

Literature cited

1. Walhagen, A.; Edholm, L.-E. J. Chromatogr. in press.
2. Walhagen, A.; Edholm, L.-E.; Heeremans, C.E.M.; Van der Hoeven,
   R.A.M.; Niessen, W.M.A.; Tjaden, U.R.; Van der Greef, J.
   J. Chromatogr. 1989, 474, 257.
3. Heeremans, C.E.M.; Van der Hoeven, R.A.M.; Niessen, W.M.A.;
   Tjaden, U.R.; Van der Greef, J. J. Chromatogr. 1989, 474, 149.
4. Zakett, D.; Kallos, G.J.; Savickas, P.J. Presented at the 32nd
   Annual Conference on Mass Spectrometry and Allied Topics. 1984,
   p. 3.
5. Niessen, W.M.A.; Van der Hoeven, R.A.M.; De Kraa, M.A.G.;
   Heeremans, C.E.M.; Tjaden, U.R.; Van der Greef, J. J. Chromatogr.
   1989, 474, 113.
6. Niessen, W.M.A.; Van der Hoeven, R.A.M.; De Kraa, M.A.G.;
   Heeremans, C.E.M.; Tjaden, U.R.; Van der Greef, J. J. Chromatogr.
   1989, 478, 325.
7. Karger, B.L.; Kirby, D.P.; Vouros, P.; Foltz, R.L.; Hidy, B.
   Anal. Chem. 1979, 51, 2324.
8. Apffel, J.A.; Brinkman, U.A.Th.; Frei, R.W. J. Chromatogr. 1984,
   312, 153.
9. Van der Greef, J.; Niessen, W.M.A.; Tjaden, U.R. J. Pharm.
   Biomed. Anal. 1988, 6, 565.

10. Verheij, E.R.; Reeuwijk, R.E.M.; LaVos, G.F.; Niessen, W.M.A.; Tjaden, U.R.; Van der Greef, J. Biomed. Environ, Mass Spectrom. 1988, 16, 393.
11. Luijten, W.; Damien, G.; Capart, J. J. Chromatogr. 1989, 474, 265.
12. Walther, H.J.; Hirter, P.; Guenat, C.; Weber, U. Adv. Mass. Spectrom. 1989, 11, 1286.
13. Verheij, E.R.; Niessen, W.M.A.; Tjaden, U.R.; Van der Greef, J. Presented at the 5th International Workshop on LC/MS, Nov. 2-4, 1988, Freiburg, F.R.G.
14. Kokkonen, P.S.; Niessen, W.M.A.; Tjaden, U.R.; Van der Greef, J. J. Chromatogr. 1989, 474, 58.
15. Van der Greef, J.; Niessen, W.M.A.; Tjaden, U.R.; J. Chromatogr. 1989, 474, 5.

RECEIVED October 24, 1989

Chapter 12

Monitoring In Vivo Cyclic Acetylation and Deacetylation of the Anticonvulsant LY201116 in Rats

Use of D_3-N-acetyl LY201116 and Liquid Chromatography/Mass Spectrometry/Thermospray Mass Spectrometry

A. P. Breau, C. J. Parli, B. D. Potts, and R. M. Goodwin

Lilly Research Laboratories, Indianapolis, IN 46285

One in vivo metabolic pathway for the anticonvulsant LY201116 (figure 1) involved a cyclic acetylation and deacetylation process. While the acetylated moiety was not pharmacologically active, the cycling of the compound provided pools of active drug. The extent of this cycling can profoundly influence the duration of pharmacological activity and pharmacokinetic half-life. In order to assess the extent of this cycling, D3- N-acetyl LY201116 was administered orally to rats at 10 mg/kg. The disappearance of D3-N-acetyl LY20116 (deacetylation) in the plasma was monitored concurrently with the appearance of N-acetyl LY201116 (de novo acetylation) and LY201116. The use of thermospray LC/MS permitted the simultaneous measurement of D3-N-acetyl LY201116 and its co-eluting non deuterated analog, N-acetyl LY201116.

LY201116 (Scheme 1 part A), a 4-aminobenzamide that is being developed as an anticonvulsant drug is N-acetylated in vivo (1). The N-acetylated metabolite, (Scheme 1 part B), is devoid of anticonvulsant activity in the rat electroshock model. Administration of N-acetyl LY201116 to mice, but not to rats, results in anticonvulsant activity concomitant with the appearance of the deacetylated metabolite, LY201116 in plasma (2-3). These observations led to the conclusion that the N-acetylation process was reversible in vivo in mice. In order to quantitate the deacetylation of exogenously administered N-acetyl LY201116 without interference from the endogenous reacetylation of the LY201116 that is formed an LC/MS method was developed. Using the mass spectrometer in multiple ion detection mode the deacetylation of an administered dose of D3-N-acetyl LY201116 (Scheme 1 part C) can be monitored without interference from the concomitant reacetylation of LY201116, because the reacetylated LY201116 formed endogenously has a pseudomolecular ion 3 mass units less than the deuterated analog. The use of deuterated analogs has been successfully employed in many metabolism studies (4) and this work represents this approach in an LC/MS application.

0097–6156/90/0420–0190$06.00/0

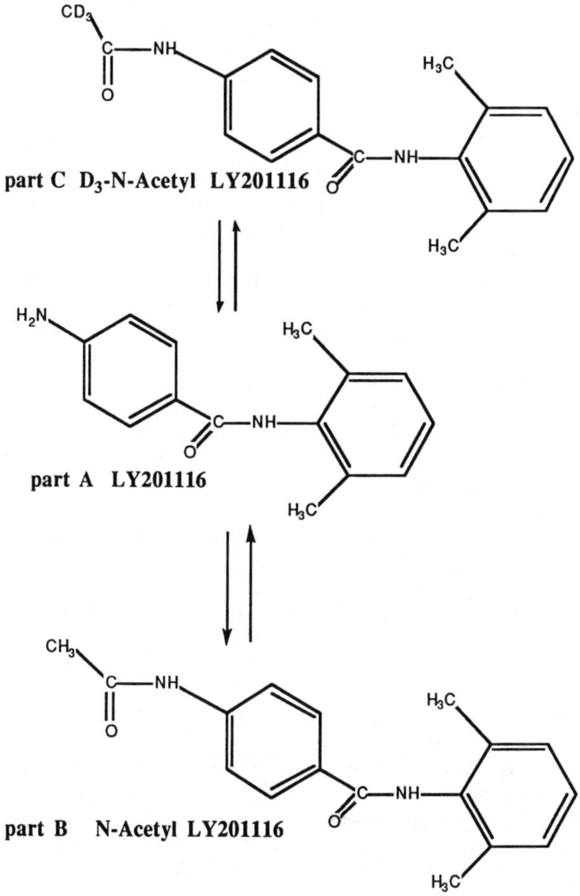

Scheme 1. The overall acetylation-deacetylattion metabolic pathway for LY201116.

Materials and Methods

Rats dosed orally at 10 mg/kg (3 rats/ sex/timepoint) were exsanguinated by cardiac puncture while under methoxyflurane anesthesia. Blood samples were collected in heparinized syringes and kept on ice until centrifugation. Blood was spun at 2000 rpm for 10 min. Plasma was removed and stored at - 30°C until analysis.

The compounds were extracted from plasma with Bond Elut CN solid phase extraction columns attached to a Vac-Elut apparatus. The columns were prewashed with 1 column volume each of methanol and water. Four hundred microliters of plasma and 50 µl of 10 µg/ml LY201409 (figure 1), as an internal standard, were put on the cartridge. The remaining column volume was filled with water and the mixture was aspirated through the column at 5-10 in. Hg vacuum. Two column volumes of water were passed through the column to eliminate water soluble plasma constituents. The compounds of interest were eluted from the column by the addition of 250 µl of HPLC mobile phase: 0.1 M ammonium acetate: methanol: acetonitrile: diethylamine (50:25:25:0.5). Extraction efficiencies for all compounds was greater than 90% (1). One hundred microliters of each plasma extract was chromatographed on a 25 cm x 4.6 mm Alltech (5 µ) R-SIL C-18 reverse-phase HPLC column. Flow rate was 1.0 ml/min. Output from the analytical column passed through a Kratos Spectraflow 783 variable wavelength U.V detector set at 270 nm. U.V. response was monitored with a Kipp and Zonen strip chart recorder. The UV response was used to quantitate LY201116 and total N-acetylated LY201116 (deuterated and non-deuterated N-acetyl LY201116).

After UV quantitation the plasma extracts were introduced into a Finnigan 4500 mass spectrometer using a Vestec thermospray interface to determine the relative amounts of deuterated versus non-deuterated N-acetyl LY201116. The limit of detection for UV analysis was 20 ng/ml for LY201116 and total N-acetyl LY201116. The LC/MS detection limit for the deuterated N-acetyl LY20116, non-deuterated N-acetyl LY201116 and internal standard was 12.5 ng on column. The LC/MS response of deuterated and non-deuterated N-acetyl LY201116 were equivalent, coinjection of equal amounts produced equal peak areas for the respective pseudomolecular ions at 283 and 286. Optimal thermospray response to the compounds of interest was obtained under the following instrumental conditions: ion source block = 300°C; probe tip = 235°C; flow = 1.0 ml/min; ionizer pressure = 2.2 x 10-5 torr. A discharge electrode was employed to increase ionization. Best thermospray results were obtained with positive ion monitoring. Most of the analyses were conducted using multiple ion detection of the following pseudomolecular ions: 299 (hydroxylated N-acetyl LY201116, #1), 241 (LY201116, #2), 283 (non-deuterated N-acetyl LY201116,#3) , 286 (D_3- N-acetyl LY201116 ,#3) and 269 (internal standard LY201409,#4) (figure 2).

Preliminary analyses were full-scan acquisitions with a mass range of 150-500 amu/2 sec. Electron multiplier voltage was set at 1500 V. Data were collected, analyzed and plotted with a Finnigan SuperIncos® data system.

Results

The LC/MS technique was able to distinguish between the deuterated N-acetyl drug and the non-deuterated N-acetyl drug by monitoring the pseudomolecular ions at m/z=286 and m/z=283 respectively (figure 2). The ratio of the intensities of the pseudomolecular ions varies from the early time point (15 minutes after administration of D_3-N-acetyl LY201116) to the later time points (8 hours after administration of D_3-N-acetyl LY201116) (figure 3). Measurement of this ratio allows one to estimate the relative deacetylation and reacetylation processes.

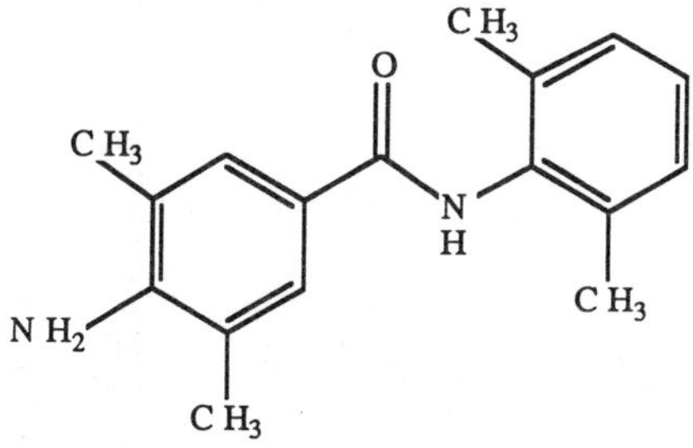

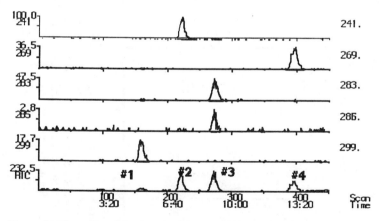

Figure 1. Structure of LY201409, the internal standard used in the HPLC analyses.

Figure 2. The selected ion monitoring trace for LC/MS analyses of hydroxylated N-acetyl LY201116 # 1, LY201116 #2, N-acetyl LY201116 and D3-Nacetyl LY201116 #3, and LY201409 (internal standard) #4.

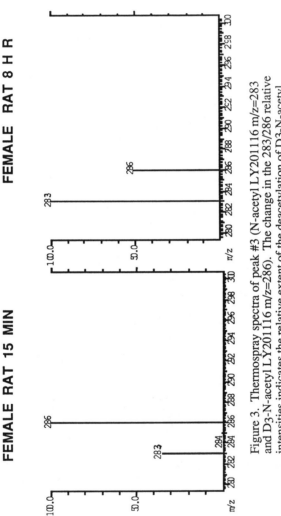

Figure 3. Thermospray spectra of peak #3 (N-acetyl LY201116 m/z=283 and D3-N-acetyl LY201116 m/z=286). The change in the 283/286 relative intensities indicates the relative extent of the deacetylation of D3-N-acetyl LY201116 versus the reacetylation of LY201116 to N-acetyl LY201116 at the different time points.

Male rats given an oral dose of D3-N-acetyl LY201116 did not have measurable (greater than 20 ng/ml) plasma levels of LY201116. However, the deuterium labelled N-acetyl LY201116 was slowly metabolized to non-deuterated N-acetyl LY201116, suggesting that deacetylation of N-acetyl LY201116 occurs at a much slower rate than reacetylation (figure 4), thus no pool of LY201116 is generated.

Sex differences were observed in male and female rats given a 10 mg/kg oral dose of D3-N-acetyl LY201116. Plasma levels of total N-acetyl LY201116 were much higher in the female rats than in the male rats (figure 4 vs figure 5).

The ratio of D3-N-acetyl LY201116 to N-acetyl LY201116 remained fairly constant in the male rats whereas this ratio decreased by an order of magnitude within 6 hours in the female rat. Plasma concentrations of LY201116 (deacetylated N-acetyl LY201116) were observed in the female rat but not in the male rat.

Discussion and Conclusions

The unique capabilities of thermospray LC/MS were used to evaluate the metabolism of LY201116. The ability to selectively monitor ions allowed the use of heavy isotope analogs to monitor two metabolic processes simultaneously. Without this approach there would not have been a method to distinguish exogenous N-acetyl LY201116 from endogenously produced N-acetyl LY201116.

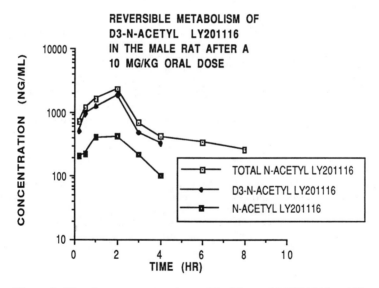

Figure 4. The plasma concentrations of D3-N-acetyl LY201116 and N-acetyl LY201116 versus time after a 10 mg/kg oral dose of D3-N-acetyl LY201116 in the male rat.

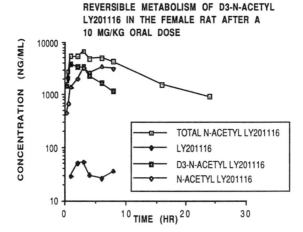

Figure 5.   The plasma concentrations of LY201116, D3-N-acetyl
LY201116 and N-acetyl LY201116 in the female rat following a 10 mg/kg
oral dose of D3-N-acetyl LY201116.

The results indicate that cyclic deacetylation of N-acetyl LY201116 and
reacetylation of LY201116 occurred in the rat.   No LY201116 was detected in the
plasma of male rats, indicating that the rats acetylated LY201116 faster than they
deacetylated N-acetyl LY201116.   The female rats deacetylated and reacetylated the
compounds to a much greater extent than did the male rats.   The fact that LY201116
was detected in the plasma of female rats may in part be the result of the higher and
prolonged plasma concentrations of N-acetyl LY201116 which could then be
deacetylated to form concentrations of LY201116 that would exceed our detectable
level of 20 ng/ml.   The greater extent of acetylation and reacetylation in the female rat
versus the male rat may also be a result of the higher concentration of N-acetyl
LY201116 in the female rat.   The underlying mechanism for these sex differences has
not yet been determined.
        Initial studies indicated that these techniques were also amenable to monitoring
the acetylation and deacetylation kinetics in mice, dogs and monkeys.   These studies
will be published in detail at a later date.   It should be noted that dogs do not acetylate
amines.   In dogs administered deuterium labelled N-acetyl LY201116, no non-
deuterated N-acetyl LY201116 was detected, further evidence that the deuterium was
not exchanged in vivo (1).

Literature Cited

1.   Parli, C.J.; Potts, B.D.; Kovach, P.M.; D.W. Robertson ; Pharmacologist
      1987, 29, 176 (abstr. .
2.   Clark, C.R. Epilepsia 1988, 29, 198-203.
3.   Parli, C.J.; Evenson, E.; Potts, B.D.; Beedle, E;.Lawson, R.; Robertson,
      D.W.; Leander,J.D. Drug Metabolism and Disposition 1988, 16, 707-711.
4.   Moor, M.J.; Rashed, M.S.; Kalhorn, T.F.; Levy, R.H.; Howald, W.N. Journal
      of Chromatography,1989, 474, 223-230.

RECEIVED November 7, 1989

ENVIRONMENTAL   ANALYSIS

Chapter 13

Analysis of Target and Nontarget Pollutants in Aqueous and Hazardous Waste Samples by Liquid Chromatography/Particle Beam Mass Spectrometry

Mark A. Brown, In Suk Kim, Fassil I. Sasinos, and Robert D. Stephens

Hazardous Materials Laboratory, California Department of Health Services, 2151 Berkeley Way, Berkeley, CA 94704

Particle Beam liquid chromatography/mass spectrometry (PB/LC/MS) based methods for the detection of target compounds daminozide, 2,4-D, Silvex, and 4-chlorobenzene sulfonic acid are presented. Separations are by reversed phase or anion exchange chromatography. Calibration curves and practical quantitation limits for electron impact (EI) and positive and negative chemical ionization (PCI and NCI) mass spectra for 21 compounds are described. EI spectra match existing libraries. Characterizing nontarget pollutants is more difficult. LC separation of nontarget compounds in aqueous leachate samples from Stringfellow and Casmalia hazardous waste sites and drinking water in California, with anion exchange chromatography PB/MS *via* EI, PCI and NCI provides only a partial characterization. Matching spectra of resolved nontarget analytes with library spectra fails apparently because they are all absent from available MS libraries.

New analytical methods are required to adequately detect and confirm the entire spectrum of environmental pollutants. Most environmental analytical methods rely on gas chromatography (GC) for resolution although the properties of most organic compounds make them unsuitable for GC separation. The group of organic compounds whose physical properties make them suitable for GC analysis is restricted to those with relatively low polarity and molecular weight. As a quantitative indication of this difference, out of currently more than six million chemicals listed in the CAS registry only approximately 123,000 mass spectra (about 2%, *via* GC inlet or direct inlet probe) have been published. Several recent studies indicate that probably most organic compounds including environmental pollutants, pharmaceuticals and agricultural chemicals and their metabolites, and many compounds of known human toxicity, all fall into the less accessible (*via* GC) but much larger region of "chemical space". HPLC is inherently suited for analyses of nonconventional pollutants that are difficult or impossible to resolve *via* conventional GC based methods because of their high polarity, low volatility or thermal instability. The coupling of HPLC with MS analysis also offers the advantage of confirmation of chemical structure, including the use of automated computer based library searches. The thermospray HPLC/MS interface has become widely adopted and has seen some application for analyses of difficult environmental analytes (1-2).

As an indication of the limitations of conventional analytical methods, one hundred and fifty chemicals were dropped from the environmental monitoring RCRA appendix VIII list in the formation of appendix IX list partially because of problems with inadequate testing capabilities for those compounds (3). In a recent survey conducted by the General Accounting Office of one hundred and five analytical laboratories across the nation, a general agreement was found among those surveyed that they could not test for these one hundred and fifty compounds (4).

0097–6156/90/0420–0198$06.00/0

Unfortunately, many of the compounds dropped from these lists come from chemical classes that are known to contain examples that are of concern to human health (Figure 1). Some, for example, show evidence of carcinogenicity and reproductive toxicity in humans and/or experimental animals (as categorized by the International Agency for Research on Cancer [IARC]); others show a large potential environmental impact and have been slated for inclusion on the EPA's ground water contamination survey (5).

Target Compound Analysis with Liquid Chromatography/Particle Beam Mass Spectrometry.

Analysis of the Plant Grown Regulator Daminozide. One example of an analytically difficult compound is daminozide (succinic acid, mono[2,2-dimethylhydrazide]), a plant growth regulator used on apples and other fruits that has come under increasing scrutiny because of concerns of its human carcinogenic potential. It is metabolically oxidized to reactive species that form adducts with DNA, and upon photochemical oxidation yields the highly toxic compound dimethylnitrosamine (6-8). Current methods do not analyze for the parent compound directly but rather use alkaline hydrolysis *in situ* to produce the unsymmetrical dimethyl hydrazine (UDMH) followed by steam distillation and derivatization of the UDMH for analysis *via* GC. This cumbersome process is further complicated by the fact that hydrolysis of daminozide to UDMH occurs naturally in fruit products (8). It is nearly impossible to extract and analyze directly from aqueous matrices because of its high polarity and water solubility. daminozide has both a basic dimethyl hydrazine and a free carboxylic acid moiety, and can be either cationic and/or anionic depending upon the pH of its solution. Although it is not a typical hazardous waste material, its properties are similar to those observed for many unknown pollutants in real world samples. Thus experience and techniques developed for its isolation, concentration and detection using ion chromatography LC/MS and solid/liquid extraction methods are directly applicable to difficult real world environmental samples.

Analysis of Chlorophenoxy Herbicides. Many important target environmental pollutants can only be detected *via* conventional GC methods by first converting them to derivatives that are less polar and more volatile, *e.g.*, the chlorophenoxy herbicides. A standard EPA method (SW-846 8150) specifies soil extraction and alkaline hydrolysis of any esters present followed by (re) esterification *via* diazomethane and detection and confirmation by GC/MS. The methylation step is required because the free carboxylic acids will not pass through conventional GC analytical columns. Reversed phase chromatography functions equally well to resolve free carboxylic acids or the corresponding esters and thus can eliminate the diazomethylation step. An interlaboratory check sample provided by the EPA of soil spiked with the chlorophenoxy acid herbicides Silvex and 2,4-D was obtained by our laboratory to demonstrate that LC/MS can offer a simpler and effective method for these compounds.

Analysis of Target Compounds in Real World Samples. Leachates from hazardous waste sites are a key source of human exposure to environmental pollutants, yet typically most compounds in this matrix are not analyzable *via* conventional methods. Thus analysis of leachates from 13 hazardous waste sites in the U.S.A. shows less than 10% of the total organic content accounted for using conventional analytical methods. The high percentage of unidentified material in this survey is considered to reflect large quantities of nonvolatile or aqueous nonextractable organic compounds, *i.e.* that portion of total organic material in the aqueous sample that is not extractable into an organic solvent under any pH conditions (9).

The Stringfellow Superfund site in California poses analytical problems similar to those encountered with most waste sites across the United States and that may be best addressed *via* LC/MS based methods. Most of the organic compounds in aqueous leachates from this site cannot be characterized by GC/MS based methods. Analysis of Stringfellow bedrock groundwater shows that only 0.78% of the total dissolved organic materials are identifiable *via* purge and trap analysis (10). These are compounds such as acetone, trichloroethylene *etc*, whose physical properties are ideally suited for GC/MS separation and confirmation. Another 33% of the dissolved organic matter is characterized as "unknown", *i.e.*, not extractable from the aqueous samples under any pH conditions and thus not analyzed *via* GC. Another 66% is 4-chlorobenzene sulfonic acid (PCBSA), an extremely polar and water soluble compound that is also not suitable for GC analysis. This compound, a waste product from DDT manufacture, is known to occur at this site because of the history of disposal of "sulfuric acid" waste from industrial DDT synthesis.

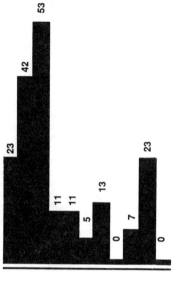

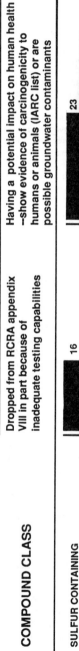

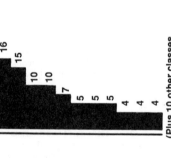

Figure 1. Frequencies of compounds, in eleven chemical classes, that have been deleted from the RCRA Appendix VIII list partially due to limitations in analytical capabilities, compared to frequencies of compounds in these same chemical classes that are considered to have a potential impact upon human health.

In another study by the same group, two major waste streams originating from Stringfellow are shown to contain fully 45% and 40% respectively of the total organic carbon as PCBSA (measured by ion chromatography and UV detection). A conventional analytical procedure using purge-and-trap analysis and acid base/neutral extraction with Stringfellow aqueous leachates shows that the majority of the total organics are not effectively fractionated but rather end up unresolved in the final aqueous remainder fraction (11). This fraction contains the highly polar and water soluble organic materials that are ideally analyzed by an HPLC based method. PB/LC/MS is clearly applicable to a problem of this nature.

Materials and Methods

Spectrometry. The Particle Beam liquid chromatography mass spectrometer consists of a Hewlett-Packard 5988a mass spectrometer equipped with a Hewlett-Packard Particle Beam HPLC interface and 1090 HPLC (Hewlett-Packard, Palo Alto, California, USA). Ionization modes include electron impact, and positive and negative chemical ionization using methane, isobutane or ammonia. LC methods are initially developed on a Hewlett-Packard 1050 HPLC equipped with a 1040 diode array detector and "Chem Station" for data acquisition. Anion exchange chromatography columns are made by SGE (Ringwood, Australia) (Model 250GL-SAX, 25 cm X 2 mm) eluting with ammonium acetate buffer and acetonitrile. Reversed phase columns (10 cm and 22 cm X 2 mm) are made by Applied Biosystems (Santa Clara, California, USA). Solvents are combinations of methanol/water or acetonitrile/water. Disposable solid phase extraction columns (reversed phase and ion exchange) are obtained from Analytichem International (Harbor City, California, USA).

Methods Development Considerations. Sensitivity limitations indicate that the direct detection of components in aqueous samples *via* PB/LC/MS is generally not possible without prior extraction and concentration. Therefore nonconventional extraction/concentration procedures are required to isolate and concentrate nonconventional pollutants. The majority of organic materials in a typical aqueous real world sample have not been extractable with organic solvents such as methylene chloride. More appropriate concentration and recovery methods include the use of lyophilization (freeze drying), various commercially available solid phase extraction and concentration techniques, as well as conventional liquid/liquid extraction procedures using organic solvents. Organic solvent liquid/solid extraction methods are still clearly the best available technique for extraction and concentration of organic analytes from solid matrices including soils. Extraction and concentration for aqueous samples uses solid phase extraction (SPE) cartridges with cation and anion exchange, reversed phase, and polar normal phase as dictated by the nature of the matrix and pollutants under analysis. Concentration of an aqueous leachate or drinking water sample by evaporation under reduced pressure or lyophilization is used to produce samples suitable for subsequent HPLC separations. Although the volatile fraction is lost, this process greatly concentrates the nonvolatile components. Solid samples such as soils, sludge or fly ash are conventionally extracted with organic solvents such as ether/aqueous methanol.

Analysis of Target Compounds. Matching EI or CI spectra and LC retention times to those obtained *via* analytical standards is exactly analogous to GC/MS methods development. Thus the major effort involves the determination of an appropriate HPLC column for any given analyte or analyte class. For example, conventional reversed phase HPLC columns are useless for extremely polar compounds such as sulfonic and certain carboxylic acids; ion exchange based columns are more appropriate.

Analysis of Nontarget Compounds, "Complete Unknowns". This is a somewhat similar process in that the retention time and the type of LC column giving the best results also yields clues as to chemical classification, *e.g.*, good retention and separation upon an anion exchange column suggests that the analytes are anionic. Confirmation information required for unknown identification is also obtained from other instrumentation including UV spectrophotometry.

Results and Discussion

Particle Beam LC/MS is Suitable for Quantitative and Qualitative Analysis of Target Compounds. PB/LC/MS is effective both as a qualitative and quantitative method for a wide range of chemical classes. Complete calibration curves and practical quantitation limits have been produced using

Table I. Practical quantitation limits (PQL, ng injected), correlation coefficients (R^2) and quadratic regression parameters (a and b) of 21 compounds with electron impact, and positive and negative chemical ionization (methane) Particle Beam mass spectrometry, direct flow injection with full scan mode

COMPOUND	Electron Impact				Positive Chemical				Negative Chemical			
	PQL	R^2	aX^2	bX	PQL	R^2	aX^2	bX	PQL	R^2	aX^2	bX
2,4-DINITROPHENOL	172	0.984	-0.002	25.3	—	—	—	—	106	0.947	-0.012	198
ACEPHATE	134	0.995	0.099	91.4	93	0.984	1.53	1,350	65.5	0.976	1.53	76.1[b]
ALDICARB	234	0.996	0.010	73.7	83.7	0.994	0.04	172	275	0.990	0.014	-4.7[b]
AZINPHOS METHYL	48.9	0.996	0.215	315	48.9	0.994	0.447	686	9.9	0.996	3.86	3,360
BROMOXYNIL	93.0	0.976	0.067	-59.0	57.2	0.987	0.289	399	10.1	0.989	21.3	8,120
DIMETHOATE	73.1	0.991	0.197	391	40.9	0.998	0.387	511	5.2	0.996	5.61	6,330
DINOSEB	—	—	—	—	712	0.939	0.004	-4.85[a]	582	0.960	0.055	-123[a]
DIQUAT	117	0.994	0.107	257	—	—	—	—	—	—	—	—

DIFENZOQUAT	56.5	0.992	0.173	570	33.1	0.999	0.176	166	—	—	—	—
ENDOSULFAN	171	0.996	0.055	2.25[b]	38.2	0.997	0.106	82.8	13.6	0.998	1.72	3,550
ETHYL PARATHION	—	—	—	—	—	—	—	—	19.8	0.998	1.76	383
FENBUTATIN OXIDE	116	0.995	0.133	336	82.9	0.983	0.059	266	80.2	0.994	0.259	184
GLYPHOSATE	778	0.947	-.0003[a]	9.47	1,040	0.980	0.0002	0.31[b]	—	—	—	—
METHIDATHION	30.6	0.999	1.09	578	28.3	0.995	0.418	1,170	4.1	0.996	3.99	2,600
METHOMYL	124	0.998	0.423	204	34.8	0.983	0.0085	38.2	—	—	—	—
MONOCROTOPHOS	159	0.988	0.095	418	19.6	0.997	0.351	550	—	—	—	—
PARAQUAT DICHLORIDE	61.3	0.994	0.136	180	209	0.998	0.010	23.5	—	—	—	—
PENDIMETHALIN	780	0.979	0.003[a]	3.16	73.4	0.992	0.265	299	16.0	0.986	1.56	-326
PROPARGITE	31.7	0.997	1.16	1,708	42.3	0.998	0.295	158	—	—	—	—
TRIFLURALIN	510	0.992	0.0040	-4.54[a]	413	0.993	0.0256	37.0[a]	—	—	—	—
ZIRAM	92.7	0.990	0.0092[a]	121	132	0.991	0.0091	69.8	—	—	—	—

Significant at $p < 0.01$ except those marked [a] $p < 0.05$ and [b] $p > 0.05$. PQL as the 95% confidence limits of the lowest amount injected (n ≥ 5).

analytical standards of 21 compounds, under various modes of ionization, that have been associated with a potential groundwater contamination in the town of Macfarland, California (Table I) (12). Figure 2 shows a representative full mass spectra and calibration curve with correlation coefficients for paraquat *via* EI ionization. Particle Beam LC/MS provides the full spectrum for these representatives of different physical classes of compounds that range from organic salts such as paraquat hydrochloride, to extremely lipophilic compounds such as fenbutatin oxide (with a molecular weight of >1,000). Not surprisingly, either positive or negative chemical ionization almost always gives greater sensitivity when compared to EI ionization. Essentially all of the calibration curves show the quadratic second order type of relationships seen in Figure 2. A reduction in the response factor (area response per unit injected) for all analytes at lower concentrations produces a loss in sensitivity at lower concentrations and effectively compresses the dynamic range.

Analysis of Daminozide, the Plant Growth Regulator, in Apple Products. Although daminozide has no retention on a reversed phase SPE cartridge or chromatography column under any pH conditions, it is well retained upon both anion exchange SPE cartridges and LC columns. Using selected ion monitoring (SIM) detection limits of 10 ng/injection on column are achieved. Figure 3 compares a SIM chromatogram of 125 ng daminozide (on a 25 cm X 2 mm SAX column eluting with ammonium acetate buffer, pH 6.0 25 mM, flow 0.25 ml/min, isobutane positive CI) to the SIM chromatogram of a sample of commercial apple juice extracted with a SAX SPE cartridge. This corresponds to a concentration of 100 ppb in this apple juice sample. Recovery under these conditions is greater than 85%.

PB/LC/MS Detection and Quantitation of Chlorophenoxy Acid Herbicides in Soil. Detection and quantitation of these target compounds is *via* PB/LC/MS with SIM and EI ionization (using 4-ions each). Figure 4 shows the PB/LC/MS chromatogram for the actual soil extract using EI/SIM. Table II compares the reference values from the interlaboratory check to the LC/MS values.

Table II. EPA laboratory evaluation data for soil spiked with the chlorophenoxy herbicides Silvex and 2,4-D comparing results from an PB/LC/MS method to EPA method SW-846 8150. The accuracy of the data is in the top 27% of all labs reporting

Analyte	Reference Value (ppm)	LC/MS value	% diff.
2,4-D	43.4	34.1	79%
Silvex	32.5	28.4	87%

These results compare very favorably with those reported from 28 other laboratories using the conventional GC/MS method. The slightly low values from the PB/LC/MS method may be in part because extraction efficiency is not corrected for. The accuracy of the PB/LC/MS based method is indicated by the fact that our values are in approximately the top 27% of all laboratories reporting in comparison with reference values. Clearly PB/LC/MS methods can yield quantitative results that are comparable with GC/MS methods, and offer very specific advantages in terms of sample preparation and simplicity of analysis.

Analysis of Stringfellow Hazardous Waste Site Groundwater. The organic extracted and purged aqueous remainder fractions were subjected to ion chromatography PB/MS analysis *via* negative CI (NCI) using isobutane as a reagent gas. The NCI spectrum of 5 µg technical PCBSA chromatographed on a 25 cm X 2 mm anion exchange column (retention time 10.2 min) shows a M⁻ ion at 192, M - 1 at 191, and a M - 35 - 1 base ion at 156 corresponding to [M - HCl]. Figure 5 shows the NCI spectrum of the major peak (retention time 8.2 min.) contained in an aqueous remainder fraction. Its good retention on anion exchange chromatography suggests that it is an organic anion, *e.g.*, a carboxylic or sulfonic acid. There is essentially no retention with reversed phase chromatography. Figure 6 shows the total ion chromatogram of a mixture of the aqueous remainder fraction (5 µl) spiked with 5 µg PCBSA standard. The two components are well separated under these conditions. New untreated samples of Stringfellow leachates were collected, portions lyophilized

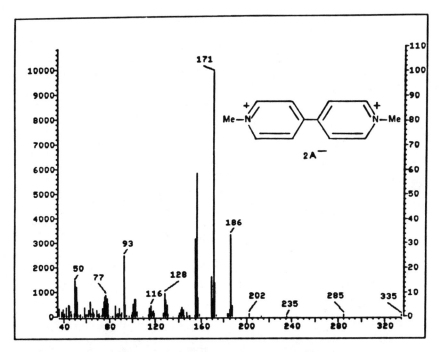

Figure 2. Paraquat dichloride (mw cation only = 186) (2.0 μg) full scan mass spectra and calibration curve (see Table I) *via* EI ionization. Continued on next page.

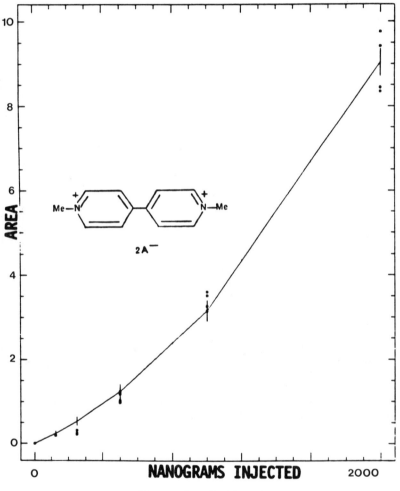

Figure 2. Continued.

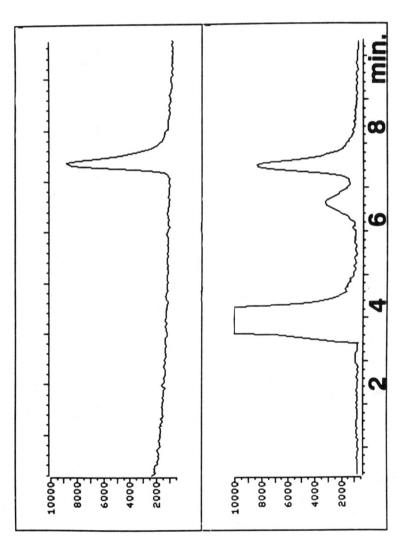

Figure 3. Anion exchange chromatogram with Particle Beam MS detection eluting with ammonium acetate buffer (selected ion monitoring, isobutane PCI) of daminozide (125 ηg) (top trace), compared to a sample of commercial apple juice extracted with a SAX SPE cartridge equivalent to 100 ppb daminozide contamination (bottom trace).

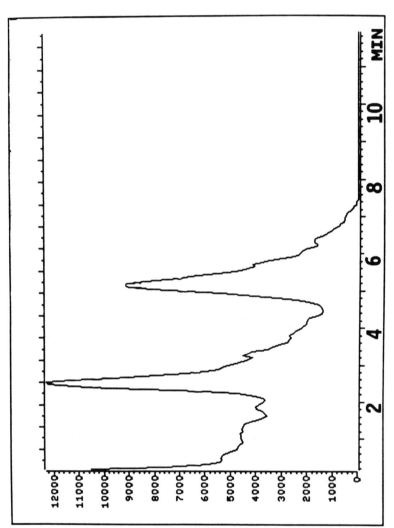

Figure 4.   Reverse phase chromatogram with Particle Beam MS detection eluting with water/methanol (selected ion monitoring, EI) of a soil extract containing the chlorophenoxy herbicides 2,4-D (R_t 2.6 min) and Silvex (R_t 5.2 min).

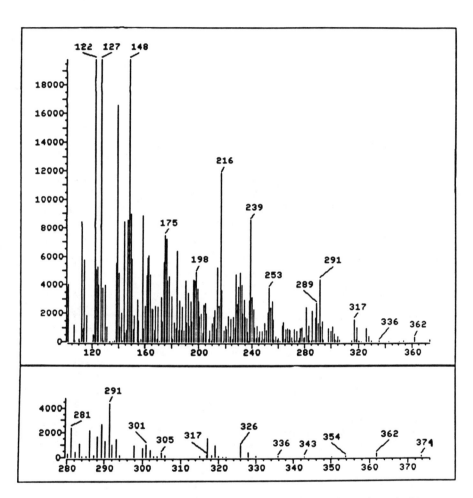

Figure 5. Particle Beam NCI (isobutane) mass spectrum of the major peak resolved by anion exchange chromatography of an aqueous remainder fraction from a Stringfellow leachate sample.

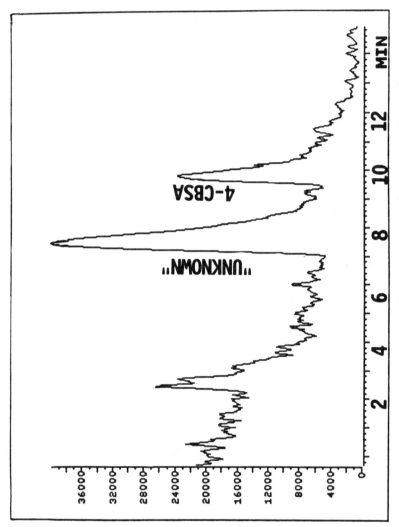

Figure 6.   Anion exchange chromatogram with Particle Beam MS detection eluting with ammonium acetate buffer (total ion, isobutane NCI) of a mixture of the Stringfellow aqueous remainder fraction (5 μl) spiked with 5 μg PCBSA standard.

and the residue subjected to anion exchange chromatography with UV (Figure 7) detection. Reversed phase and other types of commercially available HPLC columns completely fail to provide either retention or resolution of components from this concentrate. No EI spectra generated from any of these unknowns, including the PCBSA peak, are matched by EI library searches. The failure of PB/MS to provide identification of more of the components in Stringfellow samples separated *via* ion exchange chromatography indicates that additional complementary analytical techniques will be required. The UV chromatograms indicate that many of the resolved peaks, in addition to the PCBSA are aromatic compounds; this type of information compliments MS data and is an enormous aid in the consideration of tentative structural assignments for these unknown pollutants.

Anion Exchange Chromatography as a General Approach for the Resolution of Unknown Pollutants. Anion exchange chromatography PB/MS is the method of choice for the analysis of the Stringfellow leachate samples as described above, and may be the best method for resolution of components in samples from other sites. Reversed phase liquid chromatography is not useful for this type of sample nor for most actual hazardous waste leachate or aqueous samples that have been examined in this laboratory. Most of the organic materials in leachate samples appear in general to be so polar that they simply do not interact with the bonded phase of a reversed phase HPLC column.

Chromatography using a standard 22 cm reversed phase column with UV detection (210 nm) of a lyophilized ground water monitoring well sample taken from a hazardous waste site at Casmalia, California shows little or no retention. All materials elute at approximately the void volume of the column, even under solvent conditions of 98% water/2% methanol, leaving very little room for modification. Figure 8 shows the same sample chromatographed on a strong anion exchange column (SAX) (similar to the one used with the Stringfellow samples), showing resolution of a major peak at 4 minutes and at least four subsequent peaks. In a similar experiment, chromatography using a reversed phase column and UV detection of a lyophilized sample of drinking water from Santa Clara, California also shows most of the material eluting very early with little or no retention. The same sample chromatographed on the SAX column (Figure 9) shows retention and resolution of at least three peaks. The PCI (isobutane) mass spectrum of the middle peak (retention time 5.2 min) indicates that the unknown has a molecular weight of greater than 316 amu (Figure 10). An apparent periodicity of peaks separated by 14 amu suggests that the unknown may contain a long chain hydrocarbon moiety (a C_{20} carboxylate mw = 312). The full UV spectrum of this same peak (λ max 208 nm, no adsorption above 240 nm) provides the type of additional analytical data that is required for identification of unknowns; it reveals that the unknown is not aromatic, lacks extended conjugation and is consistent with an alkyl carboxylic acid. The next step with this material is to isolate sufficient material using anion exchange chromatography to obtain an FT-IR spectrum with the aim of identifying any other functional groups and confirming the presence of a free carboxylic acid.

Organic Anions may Predominate in Leachates from Hazardous Waste Sites. Clearly ion chromatography is useful for resolving the organic constituents of both leachate and drinking water samples from Stringfellow, Casmalia and Santa Clara. This suggests that a large proportion of the organic materials in these samples are organic acids. This may be due partially to the fact that many environmental chemical transformations, including microbial metabolism or photolysis, involve oxidation and hydrolysis to yield free carboxylic, sulfonic or phosphorous acids. A second possibility is that the waste materials produced by chemical manufacturing processes that are disposed of at waste sites tend to be the water soluble, anionic compounds such as the PCBSA in "sulfuric acid waste" from the chemical manufacture of DDT.

Limitations of PB/LC/MS for the Structural Elucidation of Complete Unknowns in Waste Samples. It had been anticipated that EI spectra generated by PB/LC/MS of unknown organic compounds would provide identification *via* library search algorithms. However, although good quality EI spectra have been obtained of individual peaks from the chromatography of numerous hazardous waste and drinking water samples from a wide variety of sites, only a single useful library match has been made (of diisooctyl phthalate isolated *via* reversed phase SPE from a drinking water sample). This should perhaps not be surprising; the vast majority of known organic compounds do not have EI spectra available (and needless to say the infinite number of unknown compounds have no available mass spectra).

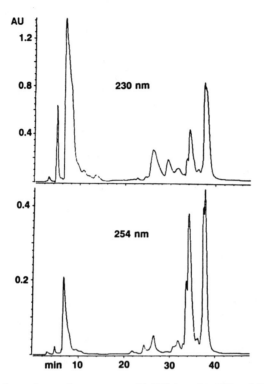

Figure 7. Anion exchange chromatogram with UV detection (230 and 254 nm) of a lyophilized extract of an aqueous leachate from the Stringfellow hazardous waste site.

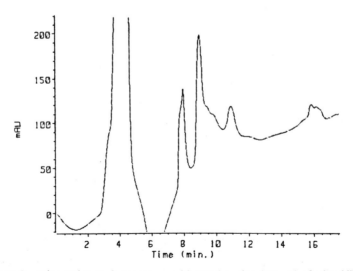

Figure 8. Anion exchange chromatogram with UV detection (210 nm) of a lyophilized ground water monitoring well sample taken from the Casmalia hazardous waste site.

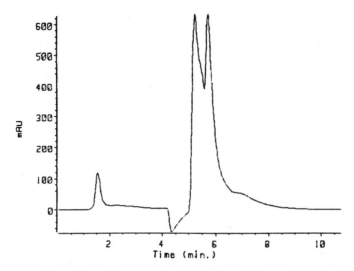

Figure 9. Anion exchange chromatogram with UV detection (210 nm) of a lyophilized sample of drinking water from Santa Clara, California.

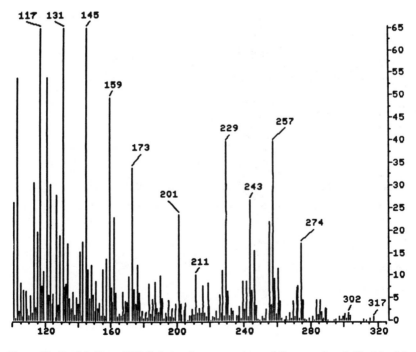

Figure 10. Particle Beam PCI (isobutane) mass spectrum of the second peak (R_t 5.2 min) from the anion exchange chromatogram of a lyophilized drinking water sample from Santa Clara shown in Figure 9.

"Unnatural Products" Chemistry. The complete identification of unknown compounds that we have successfully resolved using PB/LC/MS will clearly require additional analytical information, such as provided via liquid chromatography ICP/MS (detecting nonmetals such as chlorine and sulfur), FT-IR, UV or proton and heteroatom NMR. This situation is analogous to that of a natural products chemist faced with making a complete structural assignment of an unknown compound isolated from some matrix such as seaweed instead of a leachate from a hazardous waste site. The natural products chemist would exploit the complete array of analytical instrumentation and not attempt identification based solely upon low resolution (quadrupole) mass spectrometry.

Conclusions

Particle Beam liquid chromatography mass spectrometry has proven to be an extremely powerful confirmatory detection tool for the target compounds. Characterization of unknowns in real environmental samples, on the other hand, has proven to be more difficult. Successful concentration, separation, and EI and CI mass spectrometry, may not lead to identification. Experience has shown that rarely are either target or nontarget pollutants present in mass spectral libraries. It has become clear that a single technique such as PB/LC/MS is insufficient for the identification of unknown or nontarget pollutants. For this reason we have begun to explore the use of companion techniques including liquid chromatography ion chromatography/inductively coupled plasma/mass spectrometry.

Acknowledgments

We would like to thank EPA EMSL Las Vegas for their partial financial support for this project.

Literature Cited

1. Stephens, R. D.; Ball, N. B.; Fisher, T. S.; Roehl R.; Draper, W. M., Proc. U.S. EPA Symposium of Waste Testing and Quality Assurance, 1987, vol.1 p 15.
2. Voyksner, R. D., Anal. Chem., 1985, 57, 2600 - 2605.
3. List (Phase 1) of hazardous constituents for ground-water monitoring; final rule. Code of Federal Regulations Parts 264 and 270, 1987. Vol, 52, No. 131, 25942.
4. Hazardous Waste Report, Hazardous Waste Report, Aspen Publishers Inc., Rockville, Maryland, Feb. 17, 1986, p 9.
5. Pest. Toxic Chem. News, Food Chemical News, Inc., Washington, D.C., Jan. 23, 1985, p 16.
6. Brown, M. A.; Casida, J. E. J. Agric. Food. Chem. 1988, 36, 1064 - 1066.
7. Brown, M. A.; Casida, J. E. J. Agric. Food Chem. 1988. 36, 819 - 822.
8. Newsome, W. H. J. Agric. Food Chem., 1980, 28, 319-321.
9. Composition of Leachates from Actual Waste Sites. EPA Project Summary. 1987. Bramlett, J.; Furman, C.; Johnson, A.; Ellis, W. D.; Nelson, H.; Vick, W. H., United States Environmental Protection Agency, EPA/600/S2-87/043.
10. Stringfellow Remedial Investigation, Draft Final Report, Sections 1,2,3, Science Applications International Corp., La Jolla, California, 1987.
11. Ellis, W. D.; Bramlett, J. A.; Johnson, A. E.; McNabb, G. D.; Payne, J. R.; Harkins, P. C.; Mashni, C. I. Proc. U.S. EPA Symposium of Waste Testing and Quality Assurance, 1988, p F-46.
12. Epidemiological study of adverse health effects in children in McFarland, California, Epidemiological Studies and Surveillance Section, California Department of Health Services, Berkeley, California, Draft Report, 1988.

RECEIVED October 6, 1989

Chapter 14

Application of Combination Ion Source To Detect Environmentally Important Compounds

M. L. Vestal, D. H. Winn, C. H. Vestal, and J. G. Wilkes

Vestec Corporation, 9299 Kirby Drive, Houston, TX 77054

A new universal interface and combination
ion source is described which allows choice
of ionization modes among electron impact
(EI), chemical ionization (CI), and
Thermospray. Results obtained with this
system on a Vestec Model 201 LC-MS are
presented for some test compounds and some
environmentally important compounds on the
Appendix VIII list. The relative advantages
of the different ionization modes for
compound identification and quantitation are
discussed and data are presented on the
performance of the system.

Thermospray is well-established as a practical technique
for LC-MS interfacing and is now being used regularly in
more than 200 laboratories for a variety of
applications. (1) Despite this obvious success, its
application to environmental applications has been
limited by the fact that fragmentation is often
insufficient to allow unambiguous compound
identification. The development of the "MAGIC"
interface by Browner and coworkers (2) has demonstrated
the feasibility of obtaining EI spectra on-line with LC
separation, but the present versions of this technique
have difficulty handling the higher flow rates of
aqueous media encountered with reversed phase
chromatography using standard 4.6 mm columns. The
combined Thermospray/EI system described below was
developed in an attempt to overcome the limitations of
these earlier interfaces.

0097–6156/90/0420–0215$06.00/0

Thermospray Universal Interface

The new Thermospray "Universal Interface" was been
developed to allow HPLC to be properly coupled to
conventional EI and CI mass spectrometry. A block
diagram of the new interface is shown in Figure 1. The
LC effluent is directly coupled to a Thermospray
vaporizer in which most, but not all, of the solvent is
vaporized and the remaining unvaporized material is
carried along as an aerosol in the high velocity vapor
jet which is produced. The operation and control of the
thermospray device has been described in detail
elsewhere. (1)
     The Thermospray jet is introduced into a spray
chamber which is heated sufficiently to complete the
vaporization process. Helium is added through a gas
inlet in sufficient quantity to maintain the desired
pressure and flow rate. The fraction of the solvent
vaporized in the thermospray vaporizer and the
temperature of the desolvation chamber is adjusted so
that essentially all of the solvent is vaporized within
the desolvation region. The Thermospray system allows
very precise control of the vaporization so that all of
the solvent can be vaporized while most of even slightly
less volatile materials will be retained in the
unvaporized particles.
     After exiting the spray chamber, the aerosol
consisting of unvaporized sample particles, solvent
vapor, and inert carrier gas, passes through a condenser
and a countercurrent membrane separator where most of
the solvent vapor is removed. The resulting dry aerosol
is then transmitted to the mass spectrometer using a
momentum separator to increase the concentration of
particles relative to that of the solvent vapor and
carrier gas. The coupling between the gas diffusion
cell and the momentum separator employs a length of
teflon tubing, typically 4 mm ID and as much as several
meters long. The length of this connection is
noncritical since the dry aerosol is transmitted with no
detectable loss in sample and a negligible loss in
chromatographic fidelity.
     The apparatus in its simplest form is shown
schematically in Figure 2. The Thermospray vaporizer is
installed in the heated desolvation chamber through a
gas-tight fitting. The helium is introduced through a
second fitting and flows around the Thermospray
vaporizer and entrains the droplets produced in the
Thermospray jet. In general the carrier gas flow
required is at least equal to the vapor flow produced by
complete vaporization of the liquid input.   The heated
zone within the spray chamber should be sufficiently

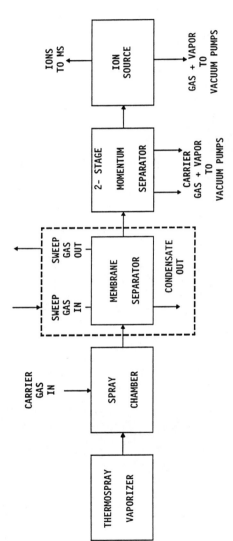

Figure 1. Block diagram of new "Universal Interface" between HPLC and EI mass spectrometry.

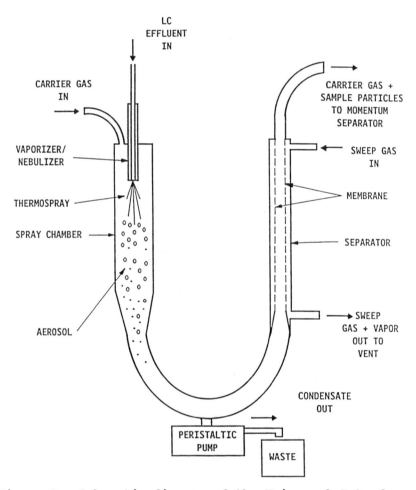

Figure 2.   Schematic diagram of the Universal Interface.

long to allow the particles to approach thermal
equilibrium with the vapor phase. The minimum
temperature for complete vaporization of the solvent is
that at which the vapor pressure of the solvent is just
greater than the partial pressure of the completely
vaporized solvent at the particular flow of liquid
employed. Ideally, the effluent from the desolvation
region consists of dry particles of unvaporized sample,
solvent vapor at a partial pressure somewhat less than
one-half of the total pressure, and the balance is the
carrier gas.

Flow velocities through the system are not
critical, but must be high enough to efficiently carry
the aerosol, but not so high as to cause extensive
turbulence. Under the correct flow conditions in which
essentially laminar flow is maintained the aerosol is
carried preferentially by the higher velocity gas stream
near the center of the tube, and the aerosol particles
can be transported for large distances with negligible
losses. Because of the very large mass of these
particles relative to the gas molecules, the diffusion
coefficients for the particles is sufficiently small
that diffusion of aerosol to the walls is very slow, and
the parabolic velocity profile provides an aerodynamic
restoring force which continually pushes particles
toward the center of the tube where the gas velocity is
highest. On the other hand, diffusion of solvent
molecules in the carrier gas is relatively rapid.
As effluent passes from the heated zone of the
spray chamber to the unheated zone of the condenser the
vapor may become supersaturated and begin to condense on
the walls. The condenser and transition region are
arranged as shown in Figure 2 so that the liquid
condensate flows under the effect of gravity to the
drain where it is pumped away to waste. A small
positive displacement pump, such as a peristaltic tubing
pump, is used to pump away the liquid without allowing a
significant amount of gas or vapor to escape.

Membrane Separator

The unique part of the Universal Interface is the
membrane separator or gas diffusion cell which allows
the solvent vapor to be efficiently removed with
essentially no loss of sample contained in the aerosol
particles. In this device the aerosol is transported
through a central channel bounded on the sides by a gas
diffusion membrane or filter medium which is in contact
with a countercurrent flow of a sweep gas. For EI mass
spectrometry helium appears to be most useful for both
the carrier and sweep gas. The properties of the

membrane appear not to be very critical, but it must be
sufficiently permeable to allow free diffusion of
carrier gas and solvent vapor while dividing the
macroscopic carrier flow from the oppositely directed
sweep flow. In this cell the concentration of vapor in
the central channel falls off exponentially, with the
value of the exponent depending on the geometry of the
cell, the effective diffusion velocity and the relative
velocities of the carrier and sweep flows. A
consequence of this exponential dependence is that the
solvent vapor transmission can be reduced as low as
required for any application merely by increasing the
length of the cell or by increasing the sweep flow. If
a certain solvent vapor removal is achieved under one
set of conditions, the square of this value can be
achieved by either doubling the length of the cell, or
by doubling the sweep gas flow.

Momentum Separator

The two stage momentum separator used in this interface
is shown schematically in Figure 3 coupled to the
combination Thermospray/EI source. This device is
conceptually similar to those used in other MAGIC (2) or
particle beam (3,4) interfaces. However, since most of
the solvent vapor is removed in the gas diffusion cell,
this separator is required primarily to remove
sufficient helium to allow the standard MS pumping
system to achieve the good vacuum required for EI
operation. The performance of this device can be
optimized much more readily for separating helium from
macroscopic particles than when copious quantities of
condensible vapors are present as in the more
conventional particle beam systems.

Combination Thermospray/CI/EI Ion Source

The beam of sample particles transmitted by the momentum
separator travels directly into the EI ion source as
shown in Figure 3. The particles impact the heated
walls of the ion source and are vaporized and ionized by
a 70 eV electron beam. The EI source is similar in
design to those used in conventional GC-MS, but the
system illustrated in Figure 3 also includes a high
pressure Thermospray source in tandem with the EI
source. With this system the Thermospray source can be
used to provide reliable molecular weight information
while the EI source provides the fragmentation needed
for structural elucidation and unambiguous
identification. Choice of EI or Thermospray operation
is selected by a single switch. Alternatively the
Thermospray source may be used as a more conventional CI

source. In this mode of operation the aerosol is coupled to the ion source using a single stage momentum separator installed in place of the thermospray vaporizer, and the exit line from the universal interface is moved from the momentum separator connected to the EI source to this single stage separator employed for CI. Reagent gas, such as methane is added directly into the high pressure ion source, and the efficient solvent removal accomplished in this arrangement allows free choice of CI reagent without significant interference from the more polar solvent vapors.

Results

Development of a commercial interface between EI-MS and HPLC using this new approach has been completed recently, and interfacing with a variety of other gas phase detectors is presently being studied. Some recent results illustrating the performance are shown in Figures 4 through 11.

Sample Transmission Efficiency

A key parameter in determining the overall performance of an LC interface is the efficiency with which sample eluted from the LC is transferred to the ion source of the mass spectrometer. Other factors such as ion source efficiency and chemical noise may determine the ultimate detection limits, but improvements in these factors generally cannot overcome the effect of sample loss on sensitivity. Results of an experiment designed to estimate sample transfer efficiency are shown in Figure 4. In this experiment peak areas of m/z 228 molecular ion from chrysene were determined for a series of flow injection inputs of chrysene in the range between 50 an 500 ng. These measurements were then repeated by loading small volumes of the same solutions on a solids probe. The probe was inserted through a 3 mm hole drilled in the side of the EI ion source opposite the inlet for the particle beam, and the bare probe was in place in the ion source during the flow injection measurements, so that sample vaporization in both experiments originated from the same region of the ion source. Sample solutions were loaded onto the probe and allowed to evaporate to dryness at room temperature and atmospheric pressure before inserting the sample into the source, to avoid sample loss from the probe. With data acquisition in progress and the LC interface operating in the normal manner with the same mobile phase used in the FIA measurements, the probe carrying the sample was then inserted into the heated source (ca. 300 C) and the m/z 228 response was integrated as before. Peak widths in both experiments were similar. From the results shown in Figure 4, we conclude that

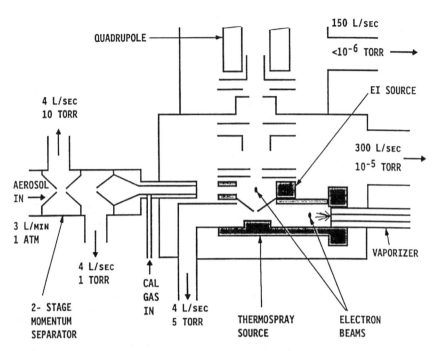

Figure 3.   Schematic diagram of combination
    Thermospray/EI ion source with two stage momentum
    separator for EI input.

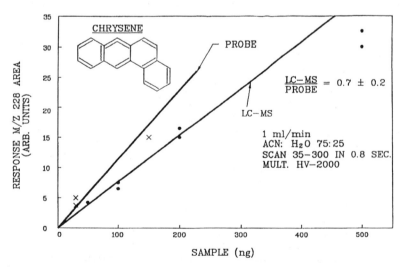

Figure 4.   Response for chrysene from flow injection
    into the LC-MS interface compared with direct probe.

under the conditions employed for these measurements about 70% of the sample is transferred from the LC to the ion source of the EI mass spectrometer.

Dependence on Liquid Flow Rate

Results for a series of injections of 2 micrograms of caffeine in 50:50 ACN:H_2O at different flow rates in the range from 0.4 to 1.5 mL/min are summarized in Figure 5. The observed peak height is proportional to flow rate, the peak width (FWHM) is inversely proportional to flow, and to within the experimental uncertainty of these measurements the peak areas are independent of flow rate. These results also provide some information on the sources of the observed sample dispersion. When the peak width is plotted as a function of the volume injected (20 microliters in this case) divided by the flow (microliters/sec) a linear relationship as shown in Figure 6 is observed. The slope of this line is a measure of dispersion in the liquid flow system as compared to ideal plug flow, and the intercept provides a measure of the flow independent dispersion in the gas phase system and the mass spectrometer. As can be seen from the Figure the dispersion in the liquid is 3.3 times the ideal and contribution from the gas phase portion is only 1.6 seconds. These results were obtained with a variable wavelength UV detector on-line which contributes significantly to the downstream band broadening. Without the UV detector similar measurements gave a slope of 1.6 and an intercept of 1.2 sec. Clearly the UV detector and its associated connections contributes somewhat more to loss of chromatographic efficiency than does the Universal Interface.

Dependence of Solvent Composition

For gradient elution the efficiency of the ideal interface should be independent of solvent composition. Some results obtained for injections of 1 microgram of caffeine into mixtures of water and acetonitrile are summarized in Figure 7. These results which show about a 30% variation in response depending on mobile phase composition were obtained without adjusting temperatures of the Thermospray vaporizer or desolvation chamber. At higher water content some sample is apparently lost in the desolvation chamber due to condensation of some droplets containing sample. This loss can apparently be minimized by optimizing the temperatures at each composition.

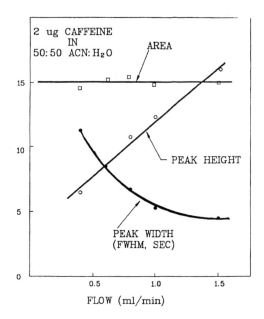

Figure 5.   Summary of measurements on the dependence on
   liquid flow rate.

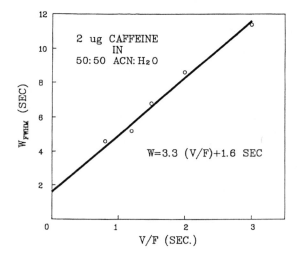

Figure 6.   Measured peak width plotted as a function of
   volume injected divided by flow rate.

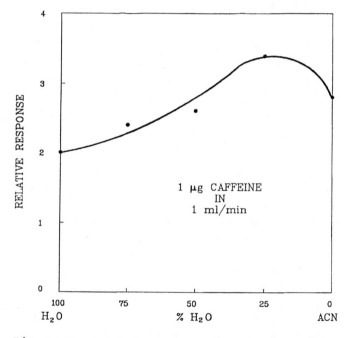

Figure 7. Dependence on solvent composition.

Solvent Removal

Figure 8 shows mass chromatograms for the molecular ion
of acetonitrile (m/z 41) and that of chrysene (m/z 228)
for a series of injections of 100 ng of chrysene into a
flow of 1 mL/min of 75:25 acetonitrile:water as the
helium sweep gas flow is reduced in steps from 10 L/min
to 4 L/min. At the higher flow the transmission
efficiency for chrysene is about 2 million times that
for the acetonitrile solvent. The corresponding
spectrum is shown in Figure 9 where the mass 41 is seen
to be small compared to the background of argon (m/z 40)
and carbon dioxide (m/z 44) due to a small air leak into
the mass spectrometer which corresponds to an analyzer
pressure of 2x10-7 torr.

Applications

This new LC-MS system is just beginning to be applied to
real environmental applications. One example of a
potential application is the hormones on the Appendix
VIII list. A total ion chromatogram for a reversed
phase separation of a mixture of hormone standards is
shown in Figure 10. These compounds yield library-
matchable spectra at the 100 ng level on column. For
example, the partially resolved doublet (peaks 2 and 3)
are easily identified as testosterone and
ethynylestradiol from their EI spectra shown in Figure
11. On the other hand, Thermospray gives only
protonated molecular ions in the positive ion mode. As
a result of the very simple spectrum Thermospray gives
much lower detection limits for these compounds (ca. 100
pg on column) while EI allows unambiguous compound
identification, but generally requires somewhat more
sample. Methane CI is intermediate between these
extremes in that it gives a simplified fragmentation
pattern at intermediate to high sensitivity. For higher
molecular weight, more complex compounds Thermospray
often gives reliable molecular weight and fragmentation
patterns that are useful for compound identification
even though it is neither as extensive or as
reproducible as that obtained by EI on smaller
molecules. In many of these cases EI fails to provide
any useful information because the compound is either
insufficiently volatile or too thermally labile to
tolerate being vaporized without undergoing extensive
pyrolysis.

Conclusion

This new LC-MS interface appears to offer substantial
promise for analysis of samples not amenable to GC-MS.
The availability of three complementary ionization

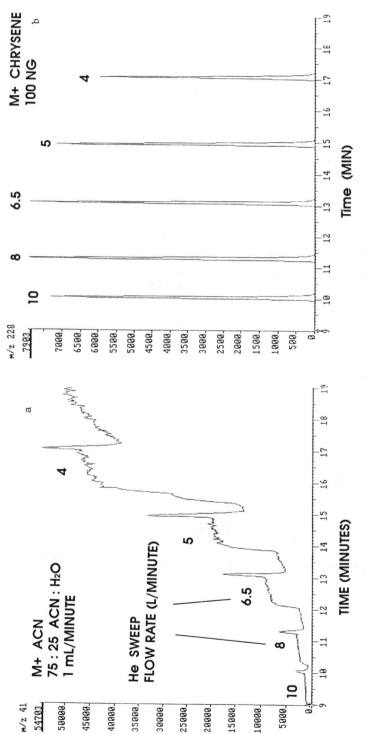

Figure 8. Mass chromatograms for (a) solvent (m/z 41 from acetonitrile) and (b) 100 ng of chrysene as functions of sweep gas flow rate.

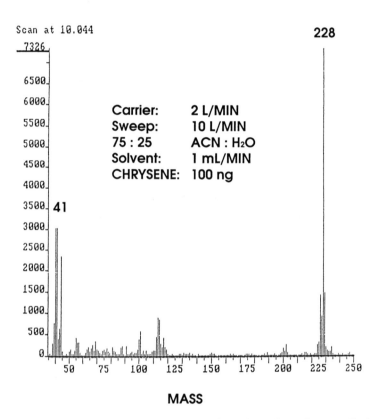

Figure 9.   Spectrum corresponding to almost complete
solvent removal at a sweep gas flow of 10 L/min.

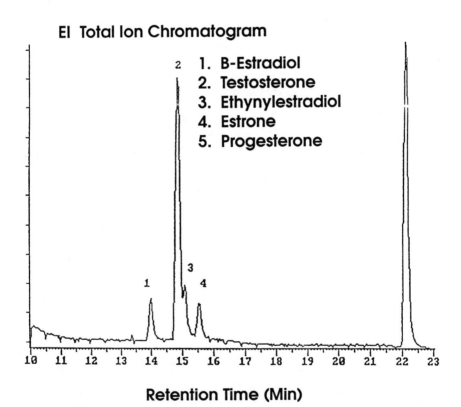

EI Total Ion Chromatogram

1. B-Estradiol
2. Testosterone
3. Ethynylestradiol
4. Estrone
5. Progesterone

Retention Time (Min)

Figure 10. Chromatogram of hormone mixture on the Appendix VIII list.

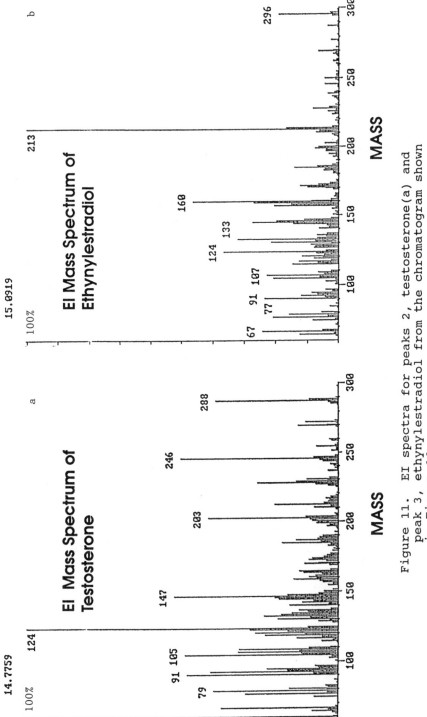

Figure 11. EI spectra for peaks 2 , testosterone(a) and
peak 3 , ethynylestradiol from the chromatogram shown
in Figure 10.

techniques on a single instrument should greatly extend the range of samples amenable to analysis and substantially improve both detection limits for targeted compounds and the reliability of unknown identification.

References

1. C. R. Blakley and M. L. Vestal, Anal. Chem. <u>55</u>, 75 (1983); M. L. Vestal and G. J. Fergusson, Anal. Chem. <u>57</u>, 2372 (1985); M. L. Vestal, Science <u>221</u>, 275 (1984).
2. R. C. Willoughby and R. F. Browner, Anal. Chem. <u>56</u>, 2626; P. C. Winler, D. B. Perkins, W. K. Williams, and R. F. Browner, Anal. Chem. <u>60</u>, 489 (1988).
3. A. Apfel, L. C. Frazier, and M. Brown, this volume. (paper 49)
4. R. C. Willoughby, E. S. Sheehan, P. E. Sanders, and M. Dilts, this volume (paper 52)

RECEIVED November 13, 1989

Chapter 15

Particle Beam Liquid Chromatography/Mass Spectrometry of Phenols and Their Sulfate and Glucuronide Conjugates

F. Reber Brown[1] and William M. Draper[2]

[1]California Public Health Foundation, 2151 Berkeley Way, Berkeley, CA 94704
[2]Hazardous Materials Laboratory, California Department of Health Services, 2151 Berkeley Way, Berkeley, CA 94704

The particle beam-liquid chromatography-mass spectrometry (PB-LC-MS) of phenols (phenol, 4-nitrophenol, and 1-naphthol) and their glucuronide and sulfate conjugates in electron impact (EI) and positive chemical ionization (PCI) is described. The compounds were separated on a strong anion exchange (SAX) HPLC column with a pH 4.5 ammonium formate-acetonitrile mobile phase. Molecular ions were not obtained for any of the conjugate structures due to decomposition, but the phenols were detected in all cases as the M^+ or $[M+H]^+$ ion. The phenol formed from each conjugate as a decomposition product could usually be identified by computerized library search. In SIM mode, limits of detection ranged from 0.25 ng for 4-nitrophenyl glucuronide to 51 ng for phenol.

In the modern analytical laboratory, gas chromatography-mass spectrometry (GC-MS) is a vital tool in the characterization and identification of unknowns. The advantages of GC-MS are accuracy in quantitation, low detection limits, tentative identification of unknowns by spectral library search, and a high degree of reliability and versatility (1). Of all the chemicals known, however, only a fraction ($\approx$20%) are amenable to analysis by GC. The remaining compounds, because of their high molecular weight, thermal instability, or ionic and/or polar character, are not suited to direct GC determination.

Some of these compounds can be analyzed by GC after derivatization to volatile species. Derivatization, however, requires extra sample handling, may give low or variable recoveries, and often requires the use of hazardous reagents. In

0097–6156/90/0420–0232$06.00/0

contrast, very few compounds are known which cannot be analyzed directly by some form of HPLC. The major limitation of HPLC has been the lack of a detector with uniformly high sensitivity. With this in mind, analytical chemists have been working for decades to couple liquid chromatographs to mass spectrometers, a combination that would theoretically allow determination of the remaining 80% of known substances.

Several interfaces have been developed for coupling HPLC's and mass spectrometers, including the moving belt, direct liquid introduction, and atmospheric pressure ionization designs. Some of these interfaces have been reviewed in the recent literature (2-5) and discussed in this symposium. The two interfaces showing the greatest potential for routine laboratory use are the thermospray (TSP) and the "monodisperse aerosol generator for introduction of chromatography" (MAGIC) to mass spectrometry, also known as a particle beam (PB) interface.

Thermospray LC-MS was developed by Vestal and co-workers and has been thoroughly reviewed (2,3,6). The PB interface (Figure 1), developed by Browner and Willoughby (7), functions as follows: The LC eluent passes through a silica capillary in a concentric pneumatic nebulizer. The nebulizer creates an aerosol of approximately uniformly sized droplets, which passes into a heated desolvation chamber held at near atmospheric pressure. Mobile phase evaporates from the droplets, forming a mixture of mobile phase vapor and analyte-containing particles which then pass through a nozzle into a two-stage momentum separator. The momentum separator pumps away the lower momentum solvent vapor and selectively transmits the higher momentum analyte particles to the MS source (8).

The PB interface accommodates common reverse and normal phase chromatographic mobile phases at flow rates up to 0.5 ml/min, and is mechanically simple, rugged, and easy to operate. Unlike the TSP interface, buffer ions are not required to effect ionization, which is instead accomplished by EI or CI in a conventional MS source (3,8). Thermospray LC-MS allows higher flow rates, but requires a special source, and generally requires a volatile buffer to ionize neutral molecules. Thermospray LC-MS provides mostly molecular weight information, whereas PB-LC-MS provides EI as well as CI spectra.

The long term objective of our work is to examine the applications of LC-MS in measuring human exposure to toxic substances. Specifically, we are investigating the direct measurement of polar and ionic metabolites of toxic compounds in the urine. Common mammalian metabolic routes include conjugation with glucuronide or sulfate moieties (9), and such conjugates are difficult to analyze by GC without extensive sample preparation (10,11). The model compounds chosen for this study and their typical parent compounds are shown in Table I.

The specific objectives of this study were to: (1) determine the general operating characteristics of the PB interface; (2) investigate the ease of adapting an existing LC-UV method to PB-LC-MS; (3) determine the capability of PB-LC-MS for the identification of polar and ionic substances.

Table I. Model Compounds Investigated by PB–LC–MS

Compound	Parent Compound/ Exposure Marker
Phenol, phenyl glucuronide,phenyl sulfate	Benzene
4-Nitrophenol, 4-nitrophenyl glucuronide, 4-nitrophenyl sulfate	Parathion
1-Naphthol, 1-naphthyl glucuronide, 1-naphthyl sulfate	Carbaryl

Experimental Section

Chemicals. Sources of the standards used are described elsewhere
(12,13). Standard solutions of the model compounds were prepared
in either distilled deionized water or pesticide grade
acetonitrile. The chromatographic mobile phase was a 2:3 (v/v)
mixture of acetonitrile and ammonium formate buffer prepared by
adjusting a 0.05 M solution of formic acid to pH 4.5 with
concentrated ammonia.

Chromatographic System. The chromatographic system used to
develop the SAX–HPLC–UV method was described previously (12).
The chromatograph used for PB–LC–MS was a Hewlett-Packard Model
1090 equipped with a ternary solvent delivery system, a variable
volume (0.1 – 25µL) sample injector, an auto sampler, and a
column bypass valve. A Hewlett-Packard Model 59980A Particle
Beam Interface was used.

Mass Spectrometer. The mass spectrometer was a Hewlett-Packard
5988A quadrupole mass spectrometer with a dual EI/CI source and
positive and negative ion detection. The system was controlled
by a Hewlett-Packard 1000 computer. The mass spectrometer was
periodically tuned manually using perfluorotributylamine (PFTBA)
on ions m/z 69, m/z 214, and m/z 502 in EI and PCI modes, and on
ions m/z 245, m/z 414, and m/z 633 in NCI mode.
    System performance was monitored daily by injecting 10 µL of
a 10 ng/µL solution of caffeine into acetonitrile using flow
injection analysis (FIA – no column in line) mode. The flow rate
was 0.4 mL/min. Five replicate injections were made, and the
baseline and mean signal-to-noise ratio (S/N) were determined.
Losses in baseline signal or S/N were indicative of problems such
as clogging or contamination of the skimmer cones in the momentum
separator, contamination of the mass spectrometer filament or
source, or electronic drift in the mass spectrometer.
    Positioning of the capillary tip in the nebulizer orifice
was critical to optimizing the system response. In order to
determine the optimum tip position, the auto sampler was set for
16 injections 18 sec apart, and the capillary was fully extended.
After each injection, the nebulizer was rotated one unit (10
units per rotation). The optimum position was that which gave
maximum ion intensity. The mass spectrometer was operated in EI
mode, with single ion monitoring (SIM) of m/z 194 with a dwell
time of 900 ms.

Flow Injection Analysis of Phenols and Conjugates. In order to determine the mode of ionization giving maximum sensitivity, FIA was used to obtain mass spectra of the individual standards of the phenols and their conjugates in both acetonitrile and acetonitrile-pH 4.5 ammonium formate (2:3, v/v) using EI, methane PCI and NCI, and isobutane PCI and NCI. Operating conditions were: flow rate, 0.4 mL/min; desolvation chamber temperature, 45°C; nebulizer pressure, 40 psi; source temperature, 250°C.

In EI mode, the scan range was m/z 50 to m/z 500 with a scan cycle time of 1.9 sec. In methane PCI and NCI, the scan range was m/z 90 to m/z 500 with a scan cycle time of 1.7 sec and the source pressure was 2×10^{-4} torr. In isobutane PCI and NCI, the scan range was from m/z 100 to m/z 500 with a scan cycle of 1.7 sec and a source pressure of 7×10^{-5} torr was used. A large increase in background precluded scanning any lower than m/z 90 or m/z 100 in methane or isobutane CI modes, respectively. The emission current was 332 µA, and the electron energy was 70 volts for EI mode and 200 volts for CI mode.

PB-HPLC-MS of Phenols and Conjugates. The SAX chromatographic separation of phenols, sulfates, and glucuronides is described in detail elsewhere (12). Briefly, a 25 cm x 4.6 mm SAX column (Supelco, Bellefonte, PA) is used with an acetonitrile-pH 4.5 ammonium formate buffer (2:3) mobile phase at a flow rate of 1.5 mL/min. Under these conditions, the nine model compounds were resolved to baseline in less than 20 minutes.

For PB-HLC-MS, the flow rate was reduced from 1.5 mL/min to 0.4 mL/min, and the nebulizer pressure was set to 50 psi. MS parameters were: ionization mode, isobutane PCI; filament current, 300 µA; electron energy, 200 V; source pressure, 8×10^{-5} torr; source temperature, 250°C.

Standards of approximately 12, 30, 60, 250, and 500 ng/µL were run in SIM mode. Ions of m/z 95.0, m/z 139.8, and m/z 144.9 were monitored for 600 msec each. These ions are characteristic ions for phenol and its conjugates, 4-nitrophenol and its conjugates, and 1-naphthol and its conjugates, respectively.

Results and Discussion

Nebulizer Optimization. The optimum position of the capillary tip in the nebulizer was determined for solvent compositions of 100%, 80%, 60%, 40%, and 20% acetonitrile in pH 4.5 ammonium formate buffer. Typical results (Figure 2) show two maxima occurring at settings of 4 and 13. For 100% acetonitrile, the maximum at 4 is greater than that at 14. In the acetonitrile-buffer systems, the maxima at 13 are slightly greater, but at this setting the capillary tip is inside the nebulizer causing spattering of the solvent and erratic baseline shifts. For this reason, the nebulizer setting of 4 was used routinely and proved stable in day to day operation.

Flow Injection Analysis of Phenols and Conjugates. A summary of the EI spectra of the phenols and their conjugates obtained in FIA mode is found in Table II. Inexplicably, phenol was not

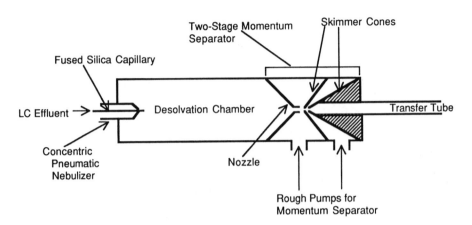

Figure 1.   Diagram of particle beam interface.

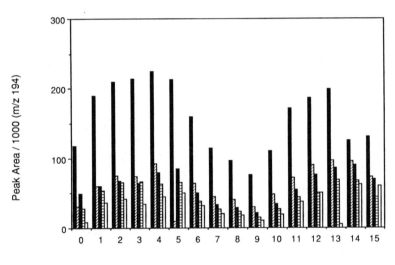

Nebulizer Position

Figure 2.   Mass spectral response vs nebulizer setting as a function of solvent composition. (■), 100% acetonitrile; (▨), 80% acetonitrile; (▤), 60% acetonitrile; (▤), 40% acetonitrile; (☐), 20% acetonitrile.

Table II. PB Electron Impact Mass Spectra for Phenols and
Conjugates [Mass(relative intensity)]

Compound	Acetonitrile	2:3 Acetonitrile-Buffer
Phenol	Not detected	Not detected
Phenyl glucuronide	173(11),114(17),94a (100),85(12),77b(16) 73(20),71(35),66(11) 65c(11)	115(9),94a(100),77b (13),71(21),66(24), 64(21),55(11)
Phenyl sulfate	94a(100),79(10),72(10), 66(21),65c(20),55(19)	116(38),115(100),109 (17),73(25),72(24), 69(15),65c(13),56 (16),55(24),54(38), 52(22)
4-Nitrophenol	139d(60),109e(34),98f (21),81(23),65g(100), 63(27),53(23)	109e(67),93f(18),81 (37),80(20),65g(100), 53(47),52(20)
4-Nitrophenyl glucuronide	139a(84),109h(50),93i (24),81(28),73(21),64 (100),63(32),53(35)	109h(78),93i(23),81 (30),65j(100),63(33) 53(54),51(32)
4-Nitrophenyl sulfate	139a(73),109h(40),93i (25),81(28),65j(100), 64(26),63(26),53(29)	109h(64),93i(21),81 (31),80(22),65j(100), 64(29),63(33),53(53), 52(29)
1-Naphthol	144d(81),116(35),115k (100),89(18),65(18),63 (28),57(13)	115k(83),77(83),61 (64),55(54),54(100), 52(50)
1-Naphthyl glucuronide	144a(100),116(42),115c (100),89(14),72(13), 66(19)	116(45),115c(100),89 (14),72(11),63(16), 57(14)
1-Naphthyl sulfate	144a(100),116(41),115c (93),89(15),72(13), 66(20)	116(39),115c(100),89 (15),64(23),63(20)

a [aglycone]+	e [M-NO]+	i [aglycone – NO₂]+
b [aglycone-OH]+	f [M-NO₂]+	j [aglycone+H-NO₂
c [aglcyone-CHO]+	g [M+H-NO₂-CHO]+	-CHO]+
d [M]+	h [aglycone-NO]+	k [M-CHO]+

detected, while 4-nitrophenol and 1-naphthol both gave prominent
molecular ions. Molecular ions for the conjugates were not
detected and in each case only the aglycone was present. This
was attributed to either hydrolysis in the desolvation chamber or
thermal decomposition in the MS source.

The results of the library search for these compounds are
summarized in Table III. Generally, the phenols were correctly
identified. The conjugates were not present in the mass spectral
library. However, they were consistently identified as their
aglycone or phenol decomposition product.

Table III. Identification of Phenols and Phenol Conjugates by
Mass Spectral Library Search

	Acetonitrile		2:3 Acetonitrile-Buffer	
	Hit#[1]	Probability Match[2]	Hit#[1]	Probability Match[2]
Phenyl glucuronide	_[4]	--	1[5]	59
Phenyl Sulfate	1[5]	73	_[3]	-
4-Nitrophenol	1	85	1	89
4-Nitrophenyl glucuronide	1[5]	96	1[5]	92
4-Nitrophenyl sulfate	1[5]	96	1[5]	91
1-Naphthol	2	64	_[4]	--
1-Naphthyl glucuronide	1[5]	89	1[5]	96
1-Naphthyl sulfate	1[5]	95	1[5]	89

[1] Position out of first five choices with 1 indicating best
match.
[2] Quality of match with 99 indicating highest quality
[3] Not identified among first five responses, but aglycone in
library.
[4] No retrievals from data base
[5] Identified as aglycone

Both signal-to-noise and absolute signal intensity were the
criteria used to select the mode of ionization for LC-MS (Table
IV). Except for the sulfates and 4-nitrophenol in PCI, there is
a decrease in S/N when the solvent is changed from acetonitrile
to acetonitrile-buffer. The opposite is observed for the
sulfates and 4-nitrophenol. In most cases, there is a parallel
loss of signal intensity when going from acetonitrile to
acetonitrile-buffer. The deleterious effects of the buffer on
S/N is least pronounced in PCI. With acetonitrile-ammonium
formate mobile phase, isobutane PCI provides optimum sensitivity
when compared to methane PCI or EI. In both methane and
isobutane NCI, peak broadening was unacceptably large, and NCI
spectra were difficult to interpret. Based on these
considerations, isobutane PCI was chosen for LC-MS studies.

The isobutane PCI spectra obtained in FIA mode are
summarized in Table V. 4-Nitrophenol and 1-naphthol were
detected as protonated molecular ions; phenol was not detected as
its protonated molecular ion is below the scan range used.

Table IV.   Signal/noise Ratio and Peak Heights
of Phenols and Conjugates

Compound (ng used)	Ionization Mode	Acetonitrile		2:3 Acetonitrile Buffer	
		S/N Ratio	Peak Height[1]	S/N Ratio	Peak Height[1]
Phenol (973)	EI	_4	_4	_4	_4
	Me-PCI[2]	39	0.48	5	0.49
	iBu-PCI[3]	_5	_5	_5	_5
Phenyl glucuronide	EI	350	39	41	0.54
(954)	Me-PCI	190	51	170	11
	iBu-PCI	_5	_5	_5	_5
Phenyl sulfate (964)	EI	68	0.89	11	0.12
	Me-PCI	43	3.3	30	0.63
	iBu-PCI	_5	_5	_5	_5
4-Nitrophenol (909)	EI	340	23	68	1.8
	Me-PCI	180	73	240	30
	iBu-PCI	150	72	240	98
4-Nitrophenyl	EI	330	19	125	2.7
glucuronide (911)	Me-PCI	140	66	82	10
	iBu-PCI	133	90	140	49
4-Nitrophenyl	EI	160	8.8	250	3.1
sulfate (964)	MePCI	69	9.4	120	13
	iBu-PCI	69	20	24	43
1-Naphthol (1091)	EI	210	5.3	7.7	0.049
	MePCI	160	10	83	2.3
	iBu-PCi	171	13	88	4.0
1-Naphthyl	EI	290	20	230	3.8
glucuronide (954)	Me-PCI	83	20	79	7.6
	iBu-PCI	110	37	100	4.0
1-Naphthyl sulfate	EI	38	3.3	42	2.7
(1027)	Me-PCI	50	4.8	69	7.8
	iBu-PCI	38	9.8	94	20

[1]   Peak Height / 10^4
[2]   Methane positive chemical ionization
[3]   Isobutane positive chemical ionization
[4]   Not analyzed - below scan range
[5]   Not detected

Again, none of the conjugated species gave molecular ions, and as before, the aglycone is detected as the [M+H]$^+$ species. The lack of molecular species for any of the conjugates in CI mode provides additional evidence for decomposition.

Table V. PB Positive Chemical Ionization Mass Spectra of Phenols and Phenol Conjugates [Mass(Relative Abundance)]

Compound	Acetonitrile	2:3 Acetonitrile-Buffer
4-Nitrophenol	181(11),140a(100)	140a(100),110b(29), 109(16)
4-Nitrophenyl glucuronide	181(10),140c(100)	140c(100),124(11)
4-Nitrophenyl sulfate	181(10),146(16), 140c(100),115(11)	168(31),140c(100), 124(14)108(17)
1-Naphthol	145a(100)	146(11)145a(100) 144(24)
1-Naphthyl glucuronide	146(12),145c(100)	146(100),145c(100)
1-Naphthyl sulfate	145c(26),140(100) 124(10),115d(43)	146(10),145c(100)

a [M+H]+	c [aglycone]+
b [M+H-NO]+	d [aglycone-CHO]+

PB-LC-MS of Phenols and Conjugates. The total ion current chromatogram and extracted ion profiles for m/z 95.0, m/z 139.8, and m/z 144.9 are shown in Figure 3. For this chromatogram, approximately 600 ng of each compound was injected. Inexplicably, phenyl glucuronide was not detected. The other peaks showed the expected retention time and elution order (13). Chromatographic efficiencies for PB-LC-MS varied from 550 theoretical plates for phenol to 8700 theoretical plates for 4-nitrophenyl sulfate. Using the HPLC-UV method, the chromatographic efficiency was somewhat better, 4300 theoretical plates for phenol and 16000 for 4-nitrophenyl sulfate. The manufacturer claims an efficiency of 7900 theoretical plates for this column. Thus, there is some added band broadening in the PB-LC-MS system.

A calibration curve for 1-naphthyl glucuronide is shown in Figure 4. For this data, the best fit straight line has a correlation coefficient of 0.989. When fitted to a quadratic equation the correlation coefficient was 1.00. Quadratic calibration curves were also seen for the other compounds in this study, and this is consistent with another report in this symposium (Brown, M. A., et al. this volume).

Instrument detection limits were estimated from the linear calibration curve using two times the average peak height of the baseline noise for each ion monitored as the minimum detectable signal. Estimated instrument limits of detection (ng) are: phenol, 51; 4-nitrophenol, 3.6; 1-naphthol, 165; 4-nitrophenyl glucuronide, 0.25; 1-naphthyl glucuronide, 5.3; phenyl sulfate, 7.7; 4-nitrophenyl sulfate, 0.29; 1-naphthyl sulfate, 3.1. For the UV detector at 254 nm, the corresponding limits of detection (ng) were: phenol, 5.0; 4-nitrophenol, 8.2; 1-naphthol, 4.0; 4-

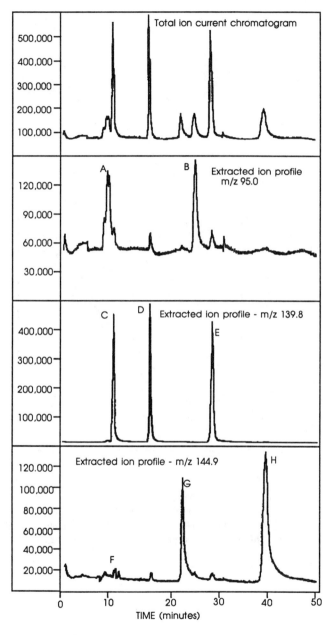

Figure 3. Total ion current chromatogram and extracted ion profiles for ions of m/z 95.0, m/z 139.8, and m/z 144.9. Peaks are identified as phenol (A), phenyl sulfate (B), 4-nitrophenol (C), 4-nitrophenyl glucuronide (D), 4-nitrophenyl sulfate (E), 1-naphthol (F), 1-naphthyl glucuronide (G), and 1-naphthyl sulfate (H).

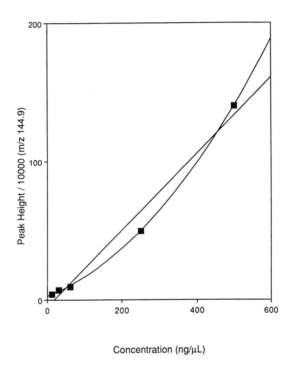

Figure 4.   Calibration curve for 1-naphthyl glucuronide
showing linear and quadratic calibration curves.

nitrophenyl glucuronide, 10; 1-naphthyl glucuronide, 14; phenyl sulfate, 85; 4-nitrophenyl sulfate, 2.8; 1-naphthyl sulfate, 7.0.

Conclusions

Flow injection analysis with the PB interface provides an effective means for introducing polar compounds to the EI source and can be viewed as an alternative to solid probe introduction. It also is a useful means for estimating the relative sensitivity in different PB-LC-MS mobile phases and ionization modes. While EI gives spectra that can be matched with online data bases, it was found to be the least sensitive operating mode. CI restricts the mass scanning range to greater than m/z 100, and may contaminate the ion source more rapidly, but gives better sensitivity and provides molecular weight information.

Adapting HPLC-UV separations to PB-LC-MS is straight forward as long as flow rate limitations are considered and volatile buffers are selected where needed. The PB-MS detector does not cause excessive peak broadening except in negative ion mode.

While glucuronide and sulfate molecular ion species are not seen in the PB mass spectra, the SAX-LC separation groups the compounds by conjugate class. Phenols elute first, followed by glucuronides, and then sulfates. Thus, tentative identification of unknown metabolites is possible based upon identification of the aglycone and the retention time window.

CA Registry Numbers

Phenol, [108-95-2]; phenyl sulfate potassium salt, [1733-88-6]; phenyl glucuronide, [17685-05-1]; 4-nitrophenol, [100-02-7]; 4-nitrophenyl sulfate potassium salt, [6217-68-1]; 4-nitrophenyl glucuronide, [10344-94-2]; 1-naphthol, [90-15-3]; 1-naphthyl sulfate potassium salt, [6295-74-5]; 1-naphthyl glucuronide, [17238-47-0].

Acknowledgements

This work was carried out at the California Department of Health Services Hazardous Materials Laboratory, and was funded by Superfund Project No. ES 04705 from the National Institute of Environmental Health Sciences of the National Institutes of Health.

Literature Cited

1. Fenselau, C. Anal. Chem. 1977, 49, 563A.
2. Arpino, Patrick J.; Guiochon, Georges. Anal. Chem. 1979, 51, 682A.
3. Covey, Thomas R.; Lee, Edgar D.; Bruins, Andries P.; Henion, Jack D. Anal. Chem. 1986, 58, 1451A.
4. Bruins, A. P. J. Chromatogr. 1985, 323, 99.
5. Arpino, Patrick J. J. Chromatogr. 1985, 323, 3.
6. Garteiz, D. A.; Vestal, M. L. Liq. Chromatogr. 1985, 3.

7. Willoughby, R. C.; Browner, R. F. Anal. Chem. 1984, 56, 2626-31.
8. Browner, R. F.; Winkler, P. C.; Perkins, D. D; Abbey, L. E. Microchemical Journal 1986, 34, 15-24.
9. Klaassen, C. D.; Amdur, M. O.; Doull, J., Eds. Toxicology: The Basic Science of Poisons; Macmillan Publishing, New York, 1986.
10. Bakke, J. E. in Bound and Conjugated Pesticide Residues; Kaufman, D. D.; Still, G. G.; Paulson, G. D.; Bandal, S. K., Eds. ACS Symposium Series No. 29; American Chemical Society: Washington, D.C., 1976; pp. 55-67.
11. Paulson, G. D. in Bound and Conjugated Pesticide Residues; Kaufman, D. D.; Still, G. G.; Paulson, G. D.; Bandal, S. K., Eds. ACS Symposium Series No. 29; American Chemical Society: Washington, D.C., 1976; pp. 86-102.
12. Brown, F. R.; Draper, W. M. J. Chromatogr. 1989, 479, 441-444.
13. Draper, W. M.; Brown, F. R.; Miille, M. J.; Bethem, R. Biomed. Environ. Mass Spectrom. 1989, 18, 767-774.

RECEIVED October 6, 1989

Chapter 16

Mass Spectrometry of the Secondary Metabolites of Benzo[a]pyrene

Ionization from Liquid Matrices

Rudolf H. Bieri and John Greaves

Virginia Institute of Marine Science, School of Marine Science, The College of William and Mary, Gloucester Point, VA 23062

Direct liquid introduction (DLI)-high performance liquid chromatography-mass spectrometry (MS), thermospray (TSP)-MS and fast atom bombardment (FAB)-MS have been used for the analysis of benzo(a)pyrene (BaP)-glucuronides and sodium salts of BaP-sulfates. Results showed that thermal decomposition, prior to ionization, predominated in DLI-MS and TSP-MS. Evidence for this included that adduct ions of BaPOH were present in the DLI-MS and TSP-MS spectra. FAB-MS did not cause thermal decomposition and gave substantial (15-100%) molecular or natriated molecular ions. Sensitivities were greater in the negative ion mode and were in the low microgram range.

During the past few years, the general sequence of events in the formation of DNA adducts with polynuclear aromatic hydrocarbons (PAH) and its possible relationship to neoplasia has been determined (1,2). Of particular importance are the biotransformations that convert PAH first to hydroxylated primary metabolites, which are then conjugated to from secondary metabolites and excreted by the organism. In the metabolic processes associated with the formation of the primary metabolites highly reactive intermediates are formed which may bind to cell macromolecules including DNA (1-3).

Analysis of PAH metabolites in samples, including those from the marine environment, has often been based on a comparison of high performance liquid chromatography (HPLC) retention volumes with those of synthetic metabolites (4-6). Derivatization of primary metabolites, followed by gas chromatographic separation and mass spectrometry (GC-MS) has also been used (7,8). Both methods are less than ideal, the HPLC lacking the necessary chromatographic resolution and specificity for complex PAH metabolite mixtures, while the GC- MS method is cumbersome, time consuming and sacrificial of information because of the need to hydrolyse the conjugated, secondary metabolites. The analysis of BaP metabolites using HPLC-MS has been reported using a direct liquid introduction (DLI) interface (9). This is the only report to the authors' knowledge on the use of soft ionization techniques for the analysis of BaP metabolites.

NOTE: This chapter is VMIS contribution No. 1559.

Due to the ubiquitous presence of PAH in the marine and estuarine environment and the observation that some fish in polluted areas have neoplastic lesions and other pathological conditions (10,11), the current research was begun to improve the information content derived from the analysis of metabolites. The eventual goal is to have a routine method to study the concentrations and the distribution of PAH metabolites in fish, so that this information can be correlated with pathological observations. This report is a part of this task and concerns the mass spectrometry of secondary metabolites.

Materials and Methods

Benzo(a)pyrene (BaP) conjugates (BaP-glucuronides = BaPglu; and sodium salts of BaP-sulfates = BaPsulfNa), were obtained from the National Cancer Institute Carcinogen Reference Repository (National Institutes of Health, Bethesda, MD). Standards were dissolved either in 0.1M aqueous ammonium acetate solution or in deionized water. For each of the three ionization techniques amounts of sample analyzed were in the low microgram range. Other chemicals were obtained from commercial suppliers and were used without further purification.

A quadrupole mass spectrometer, Model ELQ 400-2 (Extrel, Pittsburgh, PA) was used for the present studies. The lability of the compounds analysed dictated that they be introduced in a liquid matrix. The interfaces used are given below.

Direct liquid introduction (DLI) methodology has been described elsewhere (9). A thermospray probe and source (TSP, Model TS 360Q) were acquired from Vestec Corp. (Houston, TX) and later modified to include a probe tip heater incorporated in a copper block. Because of thermal contact, however, there was considerable interaction between the source temperature, the probe tip temperature and the temperature of the capillary. For this reason, attempts to optimize the ion currents by adjusting temperatures had limited success. Typical temperatures of the tip of the interface were 199- 205°C with source temperatures of 230-240°C. The two mobile phases used were 0.1M ammonium acetate and 4:1 (v/v) 0.1 M ammonium acetate: acetonitrile. The flow rate was 1.0 ml/min^{-1}. Samples were admitted to the TSP probe via a Rheodyne (Model 7125) valve.

The fast atom bombardment (FAB) source used was from Extrel. A 1:1 glycerol/thiogylcerol mixture was found to be a suitable liquid matrix. Xenon atoms were employed to bombard the target (gun conditions: 8 kV, 80-100 μA). The source heater was not used. The sample was loaded onto the FAB source either dissolved directly in the gylcerol/thioglycerol matrix, or it was deposited as an aqueous solution, dried, and then covered with glycerol/thioglycerol. The thioglycerol was necessary to dissolve the BaP conjugates while the viscosity of the glycerol extended the lifetime of the sample in the source.

Results

This paper is concerned with conjugated BaP metabolites only. Conjugated metabolites represent >90% of the metabolic products generated (12), but their lability and involatility has meant that mass spectra have been difficult to obtain. DLI-HPLC-MS, TSP-MS and FAB-MS have been used with varying degrees of success to obtain such spectra. No evidence was found of differences in the spectra that could be related to isomeric structure, despite the analysis on a number of isomers including those at the 1, 6 and 9 positions. The results reported appear to be general features of spectra of conjugated BaP metabolites. Using three different methods of sample admission and consequently different mass spectrometer sources meant that each had to be optimized for maximum sensitivity as well as for maximum signal for the respective molecular ions. Relative ionic abundances are not only a function of molecular properties, but also depend in a substantial way on source parameters (9,13,14). With these restrictions in mind, the essential characteristics of the mass spectra of BaP conjugates are given in Table I. For reasons mentioned above, the relative abundances are approximate only.

In DLI, most of the ion current is carried by the protonated (m/z 269, positive spectra) or deprotonated (m/z 267, negative spectra) aglycon or de-sulfated BaP fragment of the conjugate.

Table I. Important ions formed from BaP-glucuronides and sodium salts of BaP-sulfates using DLI-MS, TSP-MS and FAB-MS

Compound	Ionization by:	Polarity	Aglycon (rel. abund.) m/z			Molecular ions (rel. abund.)				Adduct ions present
			267	268	269	M–H	M+H	M+Na	M–Na	
BaPglu	DLI	pos.	–	–	100	–	–	–	–	–
"	"	neg.	100	–	–	–	–	–	–	$[268+CN-H]^-$, $[268+O-H]^-$, $[268+CH_2CN-H]^-$
BaPsulfNa	"	pos.	–	–	100	–	–	–	–	–
"	"	neg.	50	–	–	–	–	–	–	$[268+CN-H]^-$, $[268+O-H]^-$, $[268+CH_2CN-H]^-$
BaPglu	TSP	pos.	–	–	100	–	–	–	–	$[M+NH_4]^+$
"	"	neg.	100	–	–	5	–	–	–	$[268+Ac]^-$, $[G+Ac]^-$
BaPsulfNa	"	neg.	100	–	–	–	–	–	–	$[268+Ac]^-$
BaPglu	FAB	pos.	–	100	–	–	40	–	–	–
"	"	neg.	100	–	–	20	–	–	–	–
BaPsulfNa	"	pos.	–	25	–	–	15	100	–	$[M+Gly+H]^+$, $[M+Gly+Na]^+$, $[M+2Gly+H]^+$
"	"	neg.	100	–	–	–	–	–	35	–

268 = BaPOH
Ac = acetate
dglu = dehydrated glucuronic acid
Gly = glycerol
BaPglu = benzo(a)pyrene glucuronide
BaPsulfNa = benzo(a)pyrene sulfate, sodium salt

Jacobs et al. (15) also refer to the latter as an aglycon and for convenience this naming will be followed in this paper, although the logic is not necessarily convincing. Molecular or quasimolecular ions are ~1% or less. More interesting is the presence of ions at m/z = 283, 293 and 307 in the negative spectra, which are very likely to be the following adduct ions of the aglycon, [BaPOH−H + O]⁻, [BaPOH−H + CN]⁻ and [BaPOH−H + CH₂CN]⁻. CN⁻ and CH₂CN⁻ are derived from the acetonitrile that was present in the mobile phase and acted as a reagent gas in a chemical ionization process. The same adduct ions have previously been found in the DLI mass spectra of all hydroxy BaP (BaPOH) isomers (9). A comparison of the spectra of a BaPOH and a BaPglu is shown in Figure 1. The existence of these adduct ions for the glucuronide (Figure 1B) suggests the presence of BaPOH, formed by the thermal decomposition of the conjugated BaP metabolites prior to ionization. This is evidence that in DLI the temperature of the ion source (250°C) required to maintain the volatility of the sample is high enough to cause decomposition of the compound prior to ionization.

Another observation was that although all sulfate conjugates were admitted as their sodium salts, there was no evidence of sodium in any of the ions in the DLI mass spectra. Since these samples were introduced into the mass spectrometer in a 70:30 acetonitrile:water v/v mixture, the dissociation of the sodium salts was considered the likely reason.

TSP spectra have much in common with the DLI spectra. Again the protonated or deprotonated aglycon is the major peak in the mass spectrum and thus fits the general characteristic of glucuronic acid conjugates (16). Also present were adduct ions, this time with NH₄⁺ or acetate Ac⁻ from the 0.1M aqueous ammonium acetate in the mobile phase, which also contained 20% acetonitrile. These adducts were formed from the molecular ion, thermally derived BaPOH and a glucuronic acid derivative. The adduct ions have much in common with chemical ionization spectra (17,18). This may be expected from the dissociation of ammonium acetate and extraction of the ions from the liquid.

As with DLI, the important conclusion is that in TSP thermal decomposition at the glycosidic bond must be prevalent, although for the glucuronides the TSP spectra do have quasimolecular ions with intensities of the order of 10%. No ions in the molecular region could be detected for sulfate conjugates. Ionization of the aglycon leads to protonated ions at m/z 269 in the positive spectra, to deprotonated ions at m/z 267 in the negative mass spectra and to the formation of adducts as with the DLI spectra.

Like in DLI, there are no sodium containing ions to be found in the TSP mass spectra of the sodium salts of BaP sulfates. Dissociation of this salt is again likely to be responsible, since an 80% aqueous 0.1 M solution of ammonium acetate and 20% acetonitrile was used as the mobile phase.

Contrary to the observations for DLI and TSP, there was no detectable evidence for thermal decomposition associated with the FAB ionization process. In all compounds investigated so far, absence of thermal bond fission at the glycosidic bond in positive FAB spectra is indicated by the presence of an aglycon fragment at m/z 268 instead of the protonated aglycon at m/z 269 observed for DLI and TSP. The m/z 268 ion is the ion anticipated for fragmentation of the ionized molecule. Equally important is the fact that the relative intensities of the molecular ions are substantially larger than for TSP and DLI. For BaPsulfNa (which do not produce a molecular ion for TSP and DLI) positive FAB spectra showed the natriated adduct of the molecular ion at m/z 393 to be the base peak. In addition, fragment ions containing sodium were observed. The positive spectra of the sodium salt of the BaP sulfate conjugates also show the adducts typical of the use of glycerol as the liquid matrix (19). The positive ion spectra of BaPglu show essentially only the protonated molecular ion and the aglycon fragment. The negative ion spectra, e.g. BaP-9glu shown in Figure 2, are readily interpreted. They show only a deprotonated ion at m/z 443 for the glucuronide and a denatriated molecular ion at m/z 347 for BaPsulfNa Both types of conjugates have a fragment at m/z 267 which is the [BaPO]⁻ ion.

Discussion

Although DLI and TSP are often considered to be soft ionization mechanisms, the present study clearly proves that both BaP glucuronide and sulfate conjugates undergo thermal decomposition in both source types. Sulfate conjugates appear to be more labile than glucuronide conjugates, a fact that has received additional support from preliminary experiments using MS-MS, conducted in cooperation with VG Instruments. Because of reports which show that molecular ions of the

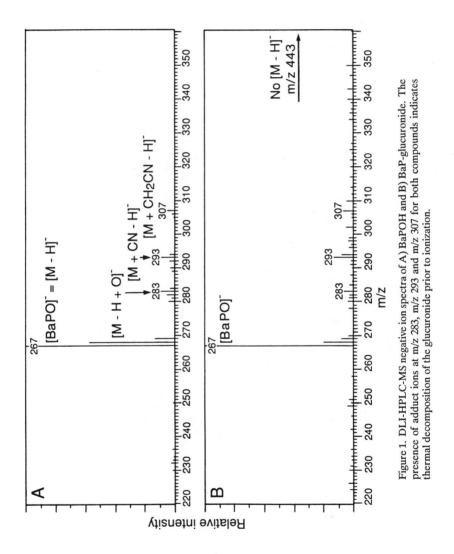

Figure 1. DLI-HPLC-MS negative ion spectra of A) BaPOH and B) BaP-glucuronide. The presence of adduct ions at m/z 283, m/z 293 and m/z 307 for both compounds indicates thermal decomposition of the glucuronide prior to ionization.

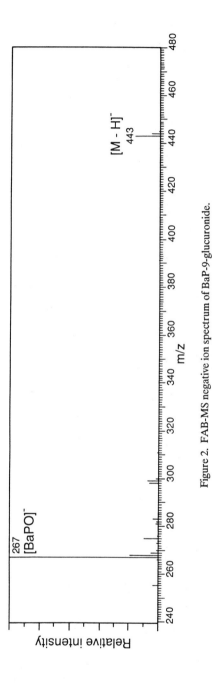

Figure 2. FAB-MS negative ion spectrum of BaP-9-glucuronide.

glucuronide and sulfate conjugates of other compounds are readily obtained by TSP, instrument performance was assessed by analysis of a number of steroid glucuronides using the same methodology. All gave spectra in which only the molecular ion was observed, (i.e. there was no cleavage of the glycosidic bond due either to thermal or ionization processes).

Evidence for thermal decomposition in DLI and TSP consists of i) the presence in the negative spectra of adduct ions of the aglycon, ii) the protonated aglycon in the positive spectra, iii) an acetate adduct of glucuronic acid and iv) the very low abundance of the quasimolecular ions (positive and negative ion spectra). Since the molecular ion is essential for the determination of the identity of the metabolites, the low intensity of these ions limits the useful sensitivity. More generally, the useful sensitivity of a technique is determined by the intensity of the smallest peak in the spectrum, needed to identify a compound of interest. For the ionization mechanisms and compounds discussed here, the sequence of useful sensitivity for negative ions is: FAB > DLI > TSP. The positive ion spectra show reduced sensitivity and often contain more fragments. The relative abundances of the latter vary, making it difficult to predict the useful sensitivities of each technique. For example, while the aglycon is normally the base peak for all three techniques, for the positive FAB spectrum of BaPsulfNa it is the natriated adduct of the molecule at m/z 393. Thus, though the sensitivity for positive ions in general may be 5% of the sensitivity for negative ions, the useful sensitivity in this case may be close to that for negative ions because of the fact that $[M + Na]^+$ is the base peak.

Although the useful sensitivity is highest for FAB, more sensitivity is expected to be required for analysis of biological samples. Optimizing the matrix composition (20), using dynamic instead of static FAB (21), and the reduction in background interferences obtained by using MS-MS are changes that may enhance sensitivity. The extraction and concentration of PAH conjugates from the bile, the biological fluid that is the major route of excretion of these compounds, can be expected to contribute to obtaining the necessary sensitivity for biological samples. Derivatization prior to FAB-MS has also been shown to enhance the sensitivity of involative compounds (22) and may therefore be applicable in the current case.

Literature Cited

1. Dipple, A.; Moschel, R. C.; Bigger, A. H. In Chemical Carcinogens; Searle, C. E., Ed.; ACS Monograph Series No. 182, 2nd edn. American Chemical Society: Washington, DC 1984; Vol. I, pp. 41-129.
2. Freudenthal, R.; Jones, P. W. Carcinogenesis - A Comprehensive Survey, Polynuclear Aromatic Hydrocarbons, Chemistry, Metabolism and Carcinogenesis; Raven Press: New York, 1976, Vol. I.
3. Gelboin, H. V. Physiological Reviews, 1980, 60, 1128-1139.
4. Gmur, D. J.; Varanasi, U. Carcinogenesis, 1982, 3 1397-1403.
5. Krahn, M. M.; Myers, M. S.; Burrows, D. G.; Malins, D. C. Xenobiotica 1984, 14, 633-646.
6. Malins, D. C.; Krahn, M. M.; Myers, M. S.; Rhodes, L. D.; Brown, D. W.; Krone, C. A.; McCain, B. B.; Chan, S.-L. Carcinogenesis 1985, 6, 1463-1469.
7. Jacob, J.; Grimmer, G. Hoppe-Seyler's Z. Physiol. Chem., 1979, 360, 1525-1534.
8. Takahashi, G.; Kinoshita, K.; Hashimoto, K.; Yasuhira, K. Cancer Research, 1979, 3, 1814-1818.
9. Bieri, R. H.; Greaves, J. Biomed. and Environ. Mass Spectrom. 1987, 14, 555-561.
10. Malins, D. C.; McCain, B. B.; Brown, D. W.; Sparks, A. K.; Hodgins, H. O.; Chan, S-L Chemical Contaminants and Abnormalities in Fish and Invertebrates from Puget Sound. National Oceanic and Atmospheric Administration, Technical Memorandum, OMP-19, 1982.
11. Hargis, W. J., Jr.; Zwerner, D. E. Mar. Environ. Res. 1988, 24, 265-270.
12. Varanasi, U.; Nishimoto, M.; Reichert, W. L.; Stein, J. E. Xenobiotica 1982, 12, 417-425.
13. Schmelzeisen-Redeker, G.; McDowall, M. A.; Giessmann, U.; Levesen, K.; Rollgen, F. W. J. Chromatogr. 1985, 323, 127-133.
14. Voyksner, R. D. Organic Mass Spectrom. 1987, 22, 513-518.

15. Jacobs, P. L.; Delbressine, P. C.; Kaspersen, F. M.; Schmeits, G. J. H. Biomed. and Environ. Mass Spectrom. 1987, 14, 689- 697.
16. Fenselau, C., In Mass Spectrom. in the Health and Life Sci.; Burlingame, A.L.; Castagnoli, N., Jr., Eds. Elsevier Science: Amstersdam, 1985, pp 321-331.
17. Rose, M. E.; Johnstone, R. A. W. Mass Spectrometry for Chemists and Biochemists, Cambridge University Press: Cambridge, 1982.
18. Fenselau, C.; Liberato, D. J.; Yergey, J. A.; Cotter, R. J. Anal. Chem. 1984, 56, 2759-2762.
19. Gower, J. L. Biomed. and Environ Mass. Spectrom. 1985, 12, 191- 196.
20. Fenselau, C.; Cotter, R. J.; Chem. Rev. 1987, 87, 501-512.
21. Caprioli, R. M., Mass Spec. Rev. 1987, 6, 237-287.
22. Wong, Q. M.; Hammargren, W. M.; Schram, K. H.; Borysko, C.; Wotring, L.; Townsend, L. B. 36th ASMS Conference on Mass Spectrometry and Allied Topics, 1988, pp 498-499.

RECEIVED October 6, 1989

Chapter 17

Anion Exchange Thermospray Tandem Mass Spectrometry of Polar Urinary Metabolites and Metabolic Conjugates

William M. Draper[1], F. Reber Brown[2], Robert Bethem[3], and Michael J. Miille[3]

[1]Hazardous Materials Laboratory, California Department of Health Services, 2151 Berkeley Way, Berkeley, CA 94704
[2]California Public Health Foundation, 2151 Berkeley Way, Berkeley, CA 94704
[3]ENSECO—California Analytical Laboratory, 2544 Industrial Boulevard, West Sacramento, CA 95691

Negative ion thermospray (TSP) tandem mass spectrometry (MS/MS) was used to determine phenols and their corresponding glucuronide and sulfate conjugates. Nine model compounds were separated to baseline on a strong anion exchange (SAX) LC column eluted with an ammonium formate buffer-acetonitrile mobile phase. Filament off TSP mass spectra provided molecular weight information for both glucuronide and sulfate conjugates (i.e., deprotonated molecule ions) as well as weak signals for sulfate hydrolysis products and dehydroglucuronic acid. TSP MS/MS provided much more structural information than TSP MS alone. Product ion spectra of the glucuronic acid conjugates revealed the aglycone mass as well as a "fingerprint" of glucuronic acid product ions. Under collisional activation the aryl sulfates fragmented efficiently to phenols detected as phenate anions. Selected reaction monitoring in LC/MS/MS allowed glucuronic- and sulfate-selective detection as well as specific detection of xenobiotics and their elaborated conjugates.

Foreign compounds absorbed by mammals are subject to a variety of metabolic processes including functionalization and conjugation, also known as Phase I and Phase II metabolism, respectively. Common Phase I transformations include oxidation, reduction, and hydrolysis while Phase II metabolism involves the biosynthesis of polar adducts (1). In general the metabolites of foreign compounds are more difficult to identify and quantitate than their parent structures due to their polarity and lower volatility.

Metabolic Conjugation of Foreign Compounds in Mammals. Evidence for Phase II metabolism was first provided by Justus von Liebig who identified hippuric acid in equine urine in 1830 (2). Soon thereafter it was demonstrated by Ure and Keller that humans treated for gout with benzoic acid also excrete hippurate in the urine. These milestones mark not only the discovery of bioconjugation, but also the beginning of the scientific study of foreign compound metabolism in mammals. By 1900 the major metabolic conjugation pathways were known and included conjugation with glycine (as in the case of hippuric acid from benzoate), sulfate, glucuronic acid, ornithine and mercapturic acid. Methylation and acetylation of foreign compounds had been reported as well.

The biochemistry of conjugation has been thoroughly reviewed in the recent publications of Caldwell (2), Hiron and Millburn (3), and a previous ACS Symposium Series monograph (4). Glucuronidation is the most versatile and quantitatively important metabolic pathway affecting a diverse group of substrates including alcohols, carboxylic acids, amines, and sulfur compounds (2,4).

Glucuronyl transferases are membrane-bound enzymes primarily associated with the endoplasmic reticulum of liver cells. Glucuronidation takes place to a lesser extent in cells of many other tissues as well. Sulfate conjugation represents the important metabolic alternative to glucuronidation being mediated by a soluble enzyme system (2). Glucuronidation is favored for lipophilic compounds while the lower molecular weight, hydrophilic xenobiotics in the cytosol are preferred substrates for sulfation.

Functionalizing and conjugating xenobiotics appears to play a role in their detoxification. Glucuronidation and sulfation usually lead to complete loss of pharmacological or pesticidal activity (5). The conjugates are highly water soluble, have reduced affinity for sensitive cellular receptors and are rapidly excreted via the kidney or bile. In rare cases Phase II bioconjugation appears to be involved in the activation of toxic compounds in the digestive and excretory organs (6).

Xenobiotics thus are cleared from the body in a variety of forms: unchanged, as Phase I metabolites, as Phase II metabolites, or products of sequential Phase I and Phase II transformations. Conjugates are the predominant excreted form for most foreign compounds (2).

Analytic Techniques for Determination of Metabolic Conjugates. The elucidation of metabolite structures was a remarkable achievement for chemists practicing before the Civil War. In spite of over 160 years of advancements in the chemical sciences, and the revolution in modern methods of instrumental analysis, structure elucidation and quantitative analysis of metabolic conjugates remain challenging. Gas-liquid chromatography (GC) or GC/MS has been used in the determination of glucuronides, but conversion to methylated, acetylated or trimethylsilylated volatile derivatives is required (7). Similarly, aryl sulfates are converted to volatile n-propyl derivatives prior to GC/MS analysis (8). Classical methods for glucuronide characterization involve enzymatic or chemical cleavage followed by identification of the aglycone (7). Similar approaches are used in sulfate conjugate identification.

Modern desorption ionization MS methods including field, plasma and laser desorption MS, fast atom bombardment (FAB) and thermospray MS eliminate the requirement of sample volatilization (9). Unlike conventional MS techniques, analytes are transported to the ion source and ionized within the condensed phase. TSP MS in particular has been applied in the determination of glucuronides (10-16), and sulfates (16,17) as well as other polar metabolites and metabolic conjugates.

Chemical Dosimetry by TSP LC/MS. One of our long-term objectives in studying TSP LC/MS is the development of chemical dosimetry based on direct determination of polar metabolites in biological fluids. Most toxic substance exposure scenarios (i.e., near hazardous waste sites) involve complex and variable mixtures of substances. Biological monitoring, where human fluids, tissues and excreta are analyzed, measures actual exposure, whereas analysis of soil, air or water can only provide an estimate of potential exposure. Exposure data forms the basis of human health risk assessment, and ultimately defines cleanup requirements at contaminated sites.

The mammalian liver through its tremendous metabolic flexibility disposes of many toxic substances, including those released from hazardous waste sites, in very few common, polar forms. In a sense we hope to exploit this capability to convert pollutants to polar, involatile forms amenable to direct TSP LC/MS determination. The development of improved means for separation, selective detection and identification of metabolic conjugates in biological fluids has applications in the dosimetry of many toxic substances. Such techniques may be useful in screening exposures to a multitude of compounds simultaneously.

Goals of this Study. The purpose of this work is to evaluate the TSP tandem mass spectrometer as a detector for glucuronide and sulfate conjugates separated on a strong anion exchange LC column. Both TSP and product ion spectra were investigated using ammonium formate buffer ionization and negative ion detection. The impact of the TSP interface on chromatographic efficiency and the capabilities of selected reaction monitoring were evaluated.

Experimental

Chemicals. Sources and preparation of chemical standards are described elsewhere (16). Chromatography solvents were commercially available, pesticide residue grade and were ultrafiltered through a Teflon filter prior to use.

<u>Thermospray MS/MS</u>. A Finnigan triple stage quadrupole (TSQ 70) instrument (San Jose, CA) equipped with a Finnigan TSP interface was operated routinely in negative ion mode with ammonium formate buffer ionization and filament off. Liberato and Yergey have referred to filament on ionization as "external" ionization, filament off ionization with or without a volatile buffer is designated "direct" thermospray ionization (<u>18</u>). A Varian Model 5000 high pressure liquid chromatgraphy (HPLC) pump with double pulse dampeners was used and samples were introduced with a 20 uL loop injector. The mobile phase was 1.5 mL/min pH 4.5 0.05 molar ammonium formate-acetonitrile (3:2, v/v). Preliminary flow injection analysis (FIA) studies (no column) reported elsewhere (<u>16</u>) used an ammonium acetate-methanol mobile phase with the filament on and both negative and positive ion detection.

Mass spectrometer settings were typically: ion source temp., 230°C; TSP aerosol temp., 265°C; vaporizer temp., 126°C. For the initial acquisition of full scan spectra, the instrument was scanned from m/z 60 to m/z 600 in 1 s. The repeller was set to 47 volts and lens settings were optimized by daily tuning with adenosine (m/z 136, m/z 268) or a polypropylene glycol mixture (acetate adduct ions at m/z 367, m/z 425, and m/z 483) or 4-chlorobenzenesulfonic acid (m/z 191 and 193) using the ammonium formate or ammonium acetate mobile phase and flow rate described above. Instrument tuning, acquisition, and data processing were carried out with Finnigan's Instrument Control Language (ICL) and ICIS data system.

Collision activated dissociation (CAD) studies used argon as a collision gas at a pressure of approximately 1.4 mtorr. Collision energies were adjusted between -9V and -40V with a goal of reducing the precursor ion intensity to less than 20% of the base peak in the product spectrum.

<u>Strong Anion Exchange HPLC Separation</u>. The strong anion exchange (SAX) LC separation of phenols and glucuronide and sulfate conjugates has been reported elsewhere (<u>19</u>). Briefly, a 5 um Supelco 4.6 mm X 25 cm LC-SAX column (Bellefonte, PA) is eluted isocratically with the ammonium formate-acetonitrile mobile phase described above. A precolumn packed with the same stationary phase and a particle filter were used. Ammonium acetate could be substituted for the formate buffer, but acetonitrile was the only organic modifier successfully used. The chromatographic conditions were optimized using an absorbance detector monitoring at 254 nm. The SAX column performance was maintained by washing the column with a phosphate buffer after use, and the column also was stored in phosphate buffer (<u>19</u>). As shown in Figure 1, phenols elute prior to glucuronides which are in turn followed by the sulfate conjugates. The mobile phase chosen was particularly advantageous for TSP and avoided postcolumn addition of a volatile TSP ionization buffer (<u>20</u>).

<u>Results and Discussion</u>

<u>Negative Ion TSP Mass Spectra</u>. Thermospray spectra for phenols revealed only deprotonated molecule ions (ArO⁻) and formate adduct ions ([M + HCOO]⁻) (Table I). 4-Nitrophenol, the strongest acid in solution, and presumably the gas phase, favored formation of ArO⁻ while phenol exhibited only the adduct ion. A deprotonated dimer anion ([2M - H])⁻ was detected in the case of 4-nitrophenol.

The spectra were similar with the LC-SAX column in line. The minor changes in ion relative intensity when FIA spectra were compared to LC/MS spectra are not uncommon in TSP mass spectrometry. TSP spectra are concentration dependent and with the column in line detector concentration is reduced due to band broadening. Thermospray mass spectrometry exhibits limited day-to-day reproducibility of ionization efficiency and fragmentation patterns, and a dependence of ion intensity on flow rate (<u>21</u>). TSP spectra also are affected by pressure, temperature, and vapor composition (<u>22,23</u>) and apparently also the design of the TSP source (<u>24</u>).

Glucuronide mass spectra also provided molecular weight information, in this case the deprotonated molecular anions were base ions for phenyl-beta-D-glucuronide and 1-naphthyl-beta-D-glucuronide (Table II). Solvolysis (or pyrolysis) was very important for 4-nitrophenyl-beta-D-glucuronide where the aglycone deprotonated molecule ion was most abundant and its formate adduct was also present. The formate adduct ion of dehydroglucuronic acid, m/z 221, was diagnostic for glucuronic acid conjugates under the thermospray conditions studied. Other glucuronic acid-derived ions including m/z 193 and m/z 175 were detected sporadically in low abundance. With filament external ionization, molecular anions (M⁻) were observed for glucuonide structures (<u>16</u>) indicating ionization by resonance electron capture. Direct thermospray ionization occurs only by proton transfer or ion attachment.

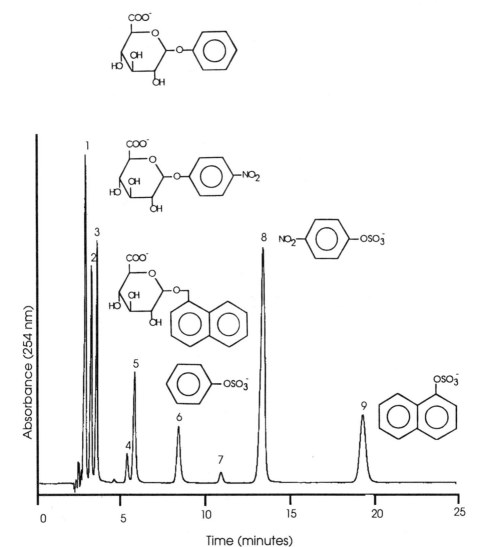

Figure 1. Strong anion exchange LC separation of phenols, aryl glucuronides and aryl sulfates using a UV absorbance detector. Compounds eluted are: 1, phenol; 2, 4-nitrophenol; 3, 1-naphthol; 4, phenyl-beta-D-glucuronide; 5, 4-nitrophenyl-beta-D-glucuronide; 6, 1- naphthyl-beta-D-glucuronide; 7, phenyl sulfate; 8, 4-nitrophenyl sulfate; 9, 1-naphthyl sulfate. (Reproduced with permission from Ref. 19. Copyright 1989 Elsevier Science Publishers B.V.)

Table I.   Thermospray Mass Spectra of Phenols
in Ammonium Formate-Acetonitrile[a]

Compound	MW	Ions (Relative Intensity, %)		
		ArO^-	$[M + HCOO]^-$	$[2M - H]^-$
phenol[b]	94	-	m/z 140(100)	-
1-naphthol	144	m/z 143(16)	m/z 189(100)	-
1-naphthol[b]	144	m/z 143(100)	m/z 189(63)	-
4-nitrophenol	139	m/z 138(100)	m/z 184(14)	m/z 277(13)
4-nitrophenol[b]	139	m/z 138(100)	m/z 184(32)	m/z 277(32)

a Flow injection (no column) spectra were recorded with filament off
and ammonium formate buffer-acetonitrile unless otherwise indicated.
The scan range for phenol was m/z 50 to m/z 150 and m/z 50 to m/z 400
for 4-nitrophenol and 1-naphthol

b HPLC-MS spectra recorded with strong anion exchange column in line
and the same mobile phase.

Table II. Thermospray Mass Spectra of Aryl Glucuronides in Ammonium Formate-Acetonitrile[a]

Compound	MW	[M − H]⁻	$[C_6H_8O_6 + HCOO]^-$	ArO⁻	Other Ions
			Ions (Relative Intensity, %)		
phenyl-b-D-glucuronide	270	m/z 269(100)	m/z 221(5.3)	–	–
phenyl-b-D-glucuronide[b]		m/z 269(100)	–		m/z 141 (20)
1-naphthyl-b-D-glucuronide	320	m/z 319(100)	m/z 221(14)	m/z 143 (5.3)	m/z 135(18), m/z 165(7.6), m/z 193(6.1)[d], m/z 259(4.6), m/z 328(4.6), m/z 387(12)
1-naphthyl-b-D-glucuronide[b]		m/z 319(100)	m/z 221(33)	–	
4-nitrophenyl-b-D-glucuronide	315	m/z 314(92)	m/z 221(27)	m/z 138(100)	m/z 175(7.6)[e], m/z 184(23)[f], m/z 284(20)[g]
4-nitrophenyl-b-D-glucuronide[b]		m/z 314(4.8)	m/z 221(19)	m/z 138(100)	m/z 175(8.0)[e], m/z 184(24)[f]

a   See footnote a in Table 1.
b   See footnote b in Table 1.
c   [M − $C_6H_8O_6$ − H]⁻
d   [glucuronic acid − H]⁻
e   [dehydroglucuronic acid − H]⁻
f   [ArOH + HCOO]⁻
g   [M − H − NO]⁻

Deprotonated molecule ions were the base peaks for phenol- and napththol sulfates. In the case of 4-nitrophenol sulfate, however, the relative intensity of the [M - H]⁻ ion was much lower (Table III). Electron-withdrawing substituents, like the p-nitro function, promote hydrolysis of sulfates, possibly explaining the susceptibility of 4- nitrophenyl sulfate to solvolysis in TSP MS. In previous external ionization TSP studies with the same source block temperature and a pH 4.5 ammonium acetate-methanol mobile phase, solvolysis of the sulfate conjugates was more prevalent (16). In fact, molecular species were not detected under these conditions. Low source block temperatures are important for obtaining molecular weight information in TSP MS (25), but this parameter was not responsible for the observed solvolysis of the aryl sulfates in this study. We propose that the cleavage of aryl sulfates in filament external ionization TSP (16) results from dissociative e⁻ capture. Electron capture chemical ionization occurs only with filament or discharge electrode external ionization. A weak signal for the deprotonated dimer was also observed for phenol sulfate in the present study.

TSP LC/MS. Negative ion LC/MS was carried out with the SAX column in line. Single ion monitoring (SIM) traces for the conjugate deprotonated molecule ions are plotted in Figure 2. The average chromatographic efficiency for the anion exchange LC/MS separation was ~6,000 theoretical plates. Using the same column, precolumn, mobile phase and flow rate, a chromatograph equipped with an absorbance detector had higher efficiency (Table IV). For the glucuronide and sulfate conjugates the average loss in chromatographic efficiency (N) was 40% when using the TSP interface.

These data demonstrate that the TSP interface, vacuum system, and ion source contribute to band broadening. While this did not effect the resolution of compounds with high capacity factors (k'), it hindered resolution of early eluting compounds like the free phenols. The total ion current (TIC) chromatogram where m/z 139 (phenol [M + HCOO]⁻), m/z 138 (4-nitrophenol [M - H]⁻), and m/z 189 (1-naphthol [M + HCOO]⁻) were monitored showed a single, broad peak at 2.58 min. In contrast, baseline separation for the phenols was accomplished with an absorbance detector (Figure 1). The band broadening associated with the TSP system emphasizes the importance of chromatographic retention and resolution of analytes for optimum specificity, particularly in conventional TSP LC/MS, and to a lesser degree in TSP LC/MS/MS.

The SIM traces plotted in Figure 2, representing 2 ug of each compound, show very little baseline noise. Instrument detection limits for these conjugates are expected to be well below 100 ng.

Product Ion Spectra. The phenols resist collisionally activated dissociation (CAD) fragmentation, even at high collision energies. In contrast formate (Table V) or acetate adduct ions (in ammonium acetate TSP) and deprotonated dimer ions dissociate, even at very low activation energies. In each case the phenate ions are base peaks in the product ion spectrum. Among the phenols, only the 4-nitrophenol [M - H]⁻ precursor ion underwent CAD fragmentation by loss of NO (Table III) and both NO and NO_2 at higher energies (16).

The sulfates as a class were fragmented by a single pathway, neutral loss of 80 mass units corresponding to loss of SO_3. Consistent with its tendency toward dissociative electron capture (or solvolysis) in filament-on TSP (16) and hydrolysis, 4-nitrophenol sulfate was most susceptible to SO_3 loss in CAD fragmentation (Table V). With a collision energy of only 14 eV, the parent ion was either absent or only a minor product ion.

Product ion spectra for glucuronide conjugates are rich in structural information in comparison to the sulfates. Diagnostic glucuronide product ions include m/z 59, m/z 85, m/z 113, and m/z 175 (Table VI) and Ref 16. The m/z 113 glucuronide-derived product anion is always prominent. Positive product ion spectra of the ammoniated glucuronide adduct exhibit a fingerprint cluster of ions as well, consisting of m/z 113, m/z 159 and m/z 193 or m/z 194 (16).

The aglycone depronated molecule anion is a prominent product ion at low collision energies, and with increasing collision energies glucuronide-derived ions diminish leaving only the aglycone signal. In a previous study of glucuronide CAD mass spectra, 20 ug samples were introduced by flow injection (16). The diagnostic spectral features, however, are also detected with 2 ug samples in LC/MS spectra where band broadening further lowers detector concentration.

Most of the fingerprint product ions originate solely from CAD fragmentation except for the ubiquitous dehydroglucuronic acid deprotonated molecule ion, m/z 175. This ion is observed in TSP spectra, particularly in cases like 4-nitrophenyl-beta-D-glucuronide where facile solvolysis occurs. The conjugate cleavage mechanism occurring both in the thermospray source and the

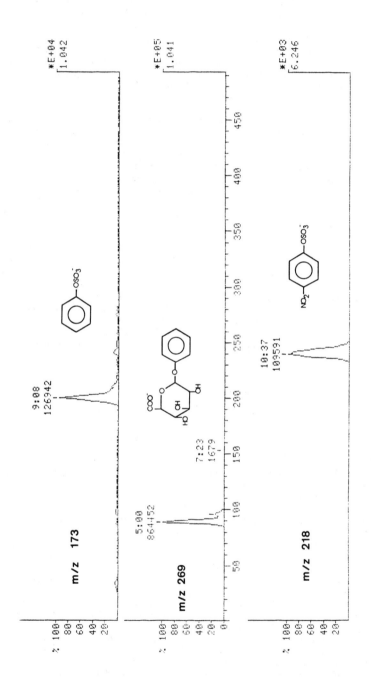

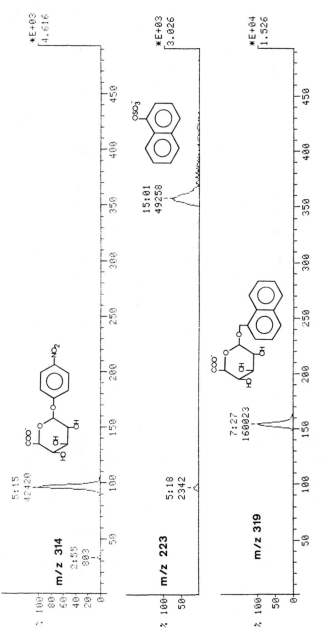

Figure 2. TSP SIM traces for metabolic conjugates eluted from a strong anion exchange column. Deprotonated molecule ions are plotted as follows: m/z 173, phenyl sulfate; m/z 269, phenyl-beta-D- glucuronide; m/z 218, 4-nitrophenyl-beta-D-glucuronide; m/z 314, 4-nitrophenyl-beta-D-glucuronide; m/z 223, 1-naphthyl sulfate; m/z 319, 1-naphthyl-beta-D-glucuronide.

Table III. Thermospray Mass Spectra of Aryl Sulfates[a]

Compound	MW[b]	Ions (Relative Intensity, %)			
		$AroSO_3^-$	$[ArOH + HCOO]^-$	ArO^-	Other Ions
phenol sulfate, K+ salt	173	m/z 173(100)	-	-	m/z 347(2.5)[c]
4-nitrophenol sulfate, K+ salt	218	-	m/z 184(13)	m/z 138(100)	m/z 277(21)[d]
"[e]		m/z 218(42)	m/z 184(33)	m/z 138(100)	m/z 277(4)[d]
1-naphthol sulfate, K+ salt	223	m/z 223(100)	m/z 189(7.6)	-	-
"[e]		m/z 223(100)	m/z 189(2.4)	-	-

a Flow injection (no column) with 1.5 mL/min of 0.05 molar ammonium formate-acetonitrile (3:2, v/v)
b Anion mass
c $[2\ AroSO_3 + H]^-$
d $[2\ ArOH - H]^-$
e HPLC MS spectrum with strong anion exchange column in line

Table IV. Chromatographic Efficiency Determined for LC/UV and TSP/LC/MS Separation of Glucuronides and Sulfates on a Strong Anion Exchange Column

Compound	LC/UV		TSP/LC/MS		
	t_R (min)	N^a	Ion (m/z)	t_R (min)	N
phenyl glucuronide	5.37	7,100	269	5.00	2,900
4-nitrophenyl glucuronide	5.79	8,300	314	5.15	2,700
1-naphthyl glucuronide	7.15	5,600	319	7.27	7,000
phenyl sulfate	11.0	17,000	173	9.08	12,000
4-nitrophenyl sulfate	13.4	16,000	218	10.4	4,300
1-naphthyl sulfate	19.5	12,000	233	15.0	7,500

a N = chromatographic efficiency calculated from peak width at half height using the following equation: $N = 5.54 \ (\ t_R \ / \ w_{1/2} \ )^2$

Table V.   CAD spectra of phenols and aryl sulfates [a]

Compound	Precursor Ion	CAD Product Ions		
		Proposed Ion	m/z	RI (%)[b]
1-naphthol	[M + HCOO]⁻	HCOO⁻	45	<5
		[M − H]⁻	143	100
		[M + HCOO]⁻	189	<5
4-nitrophenol	[M − H]⁻	[M − H − NO]⁻	108	17
		[M − H]⁻	138	100
phenol sulfate	[M − H]⁻	[M − H − SO₃]⁻	93	100
		[M − H]⁻	173	67
1-naphthyl sulfate	[M − H]⁻	[M − H − SO₃]⁻	143	100
		[M − H]⁻	223	43 − 79
4-nitrophenyl sulfate	[M − H]⁻	[M − H − SO₃]⁻	138	100
		[M − H]⁻	218	0 − 24

a   The collision energy was 14 eV in each except 1-naphthol where the
    collision energy was 9 eV.
b   Relative intensity

Table VI. CAD mass spectra of glucuronic acid conjugates [a]

Compound	Precursor Ion	CAD Product Ion (RI)[b]
phenyl-b-D-glucuronide	[M − H]−	59(7.6), 75(<5), 85(11), 93(14)[c], 99(17), 113(57), 117(13), 175(10)[d], 269(100)[e]
4-nitrophenyl-b-D-glucuronide	[M − H]−	75(13), 85(11), 113(16), 138(100)[c], 175(22)[d]

a LC/MS spectrum for 2 ug with the SAX column in line
b Relative intensity (%)
c [ArO]−
d [C$_6$H$_7$O$_6$]−
e [M − H]−

argon collision chamber appears to be predominantly beta-elimination giving dehydroglucuronic acid and phenol directly. True solvolysis, involving attack by water, gives glucuronic acid instead, which was also detected in the 1-naphthyl-beta-D-glucuronide TSP spectrum, m/z 193 (Table II).

Selected Reaction Monitoring (SRM).   One of the greatest strengths of the tandem mass spectrometer is its versatility in monitoring chromatographic eluents. Conventional LC/MS is accomplished at the first quadrupole (Q_1) by either mass scanning or single ion monitoring of positive or negative ions. The instrument used in this work allows pulsed positive and negative ion detection, a capability that provides complementary information in the case of glucuronide conjugates (16).

Ions transmitted to the second quadrupole, so-called precursor ions, are accelerated under the influence of an electrical field, and collide with argon atoms. The severity of the collisions, the amount of energy imparted and the resulting fragmentation is determined by collision gas pressure and mass and the applied potential. Fragments formed on collisional activation dissociation are transmitted to the third stage quadrupole, Q_3, where the product ion spectrum can be recorded in its entirety in so-called CAD product scanning. In a variation of single ion monitoring in TSP MS, the scanning in Q_1 is linked to that of Q_3 achieving "neutral loss" scanning.

Examples of neutral loss SRM are shown in Figure 3. As described above the aryl sulfates fragment almost exclusively by neutral loss of SO_3. Thus, simultaneous control of the Q_1 and Q_3 mass filters to pass ions m/z X and m/z (X - 80), respectively, allows sulfate conjugate- selective detection. Accordingly, phenol sulfate (m/z 173 --> m/z 93), 4-nitrophenol sulfate (m/z 218 --> m/z 138), and naphthol sulfate (m/z 223 --> m/z 143) are monitored (Figure 3). At this time we have not applied this technique to sufficient numbers of biological samples, and cannot yet refer to the technique as sulfate conjugate "specific".

An analogous process is used for selective detection of glucuronic acid conjugates. CAD fragmentation of the glucuronide molecule anion is characterized by two processes: 1) beta-elimination yielding the aglycone (detected as ArO⁻) and dehydroglucuronic acid (detected as its deprotonated molecule ion, m/z 175); and 2) further fragmentation of dehydroglucuronic acid. Universal and selective glucuronide detection is achieved by monitoring the neutral loss of 176 mass units from the deprotonated molecule ion to give the aglycone deprotonated molecule anion. Phenyl-, 4-nitrophenyl-, and 1-naphthyl- beta-D-glucuronides are thus detected as 269 --> 93, 314 --> 138, and 319 --> 143 neutral losses, respectively (Figure 3).

The tandem mass spectrometer also can operate as a relatively non specific detector by, for example, monitoring the neutral loss of 46 mass units (HCOOH) when using ammonium formate buffer ionization or 60 mass units (CH_3COOH) when using acetate buffer ionization. In Figure 3, naphthol is detected in such a manner by monitoring the 189 --> 143 neutral loss.

SRM provides additional selectivity in detecting an individual xenobiotic and its related metabolites. A moiety or substructure in the xenobiotic is tagged by its novel CAD fragmentation. In the case of 4-nitrophenol, the neutral loss of 30 mass units (NO) labels the nitrophenyl moiety (actually the NO_2 substituent). With Q_1 locked at m/z 138 and Q_3 locked at m/z 108 we detect 4-nitrophenol and its glucuronide and sulfate conjugates (Figure 3). Presumably, other metabolite structures in which the nitrophenol moiety is retained would be detected similarly.

In practical application scanning can be manipulated on-the-fly within a chromatographic separation to obtain maximum information. In metabolism studies or as a chemical dosimeter, the structural feature of the parent compound and its unique neutral loss occurring on collisional activation, marks the metabolite. The mass of the metabolite is then obtained from the TSP mass spectrum at Q_1 and the product ion spectrum of the metabolite molecule ion is obtained by product ion scanning. Recent publications have discussed additional applications of both tandem mass spectrometry (26, 27) and thermospray tandem mass spectrometry (28) in metabolite structure elucidation.

Conclusion

As a soft ionization technique thermospray mass spectrometry often provides little fragmentation of molecular species. More importantly from the viewpoint of a metabolism chemist, thermospray accomplishes desorption ionization of extremely low vapor pressure analytes including intact glucuronide and sulfate conjugates. As demonstrated in the chapter by Brown and Draper in this proceedings, particle beam mass spectrometry does not have this capability.

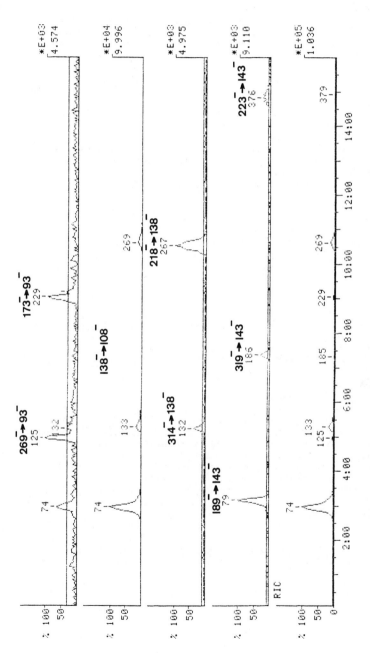

Figure 3. Selected reaction monitoring of aryl glucuronides and sulfates eluted from a strong anion exchange column. Glucuronides are detected by neutral loss of 176 mass units as in phenyl-beta-D-glucuronide (269⁻ --> 93⁻), 4-nitrophenyl-beta-D-glucuronide (314⁻ --> 138⁻), and 1-naphthyl-beta-D-glucuronide (319⁻ --> 143⁻). Sulfate conjugates are detected by neutral loss of SO₃ as in phenyl sulfate (173⁻ --> 93⁻), 4-nitrophenyl sulfate (218⁻ --> 138⁻) and 1-naphthyl sulfate (223⁻ --> 143⁻). Compounds with a 4-nitrophenol moiety are detected with Q₁ at m/z 138 and Q₃ at m/z 108: scan 74, 4-nitrophenol; scan 133, 4-nitrophenyl-beta-D-glucuronide; and scan 269, 4- nitrophenyl sulfate.

Separation of metabolite conjugates with the strong anion exchange HPLC column is extremely well suited to thermospray mass spectrometry. This LC/MS application has not been previously reported in the literature. The ammonium formate/acetonitrile or ammonium acetate/acetonitrile mobile phases supply the volatile ions required for TSP ionization.

The LC/UV separation was readily adapted to TSP LC/MS with some loss in chromatographic efficiently. The band broadening encountered restricted the resolution of low k' compounds, but was inconsequential in the baseline separation of the the aryl glucuronide and aryl sulfate compounds studied. Losses in efficiency observed in TSP LC/MS stress the importance of chromatographic retention and phase selectivity in obtaining high quality, interference free LC/MS spectra.

Product ion spectra provide a fingerprint of diagnostic ions as well as the aglycone mass for glucuronide conjugates. The prominence of these two spectral features can be controlled by the collision energy. The sulfate conjugates fragment by a single major process in the collision chamber, neutral loss of SO_3. Through selected reaction monitoring the tandem mass spectrometer affords glucuronide- or sulfate specificity, a capability of obvious importance in the elucidation of metabolite structures. The tandem mass spectrometer's ability to selectively detect substructures of a xenobiotic (i.e., the nitrophenate moiety by the neutral loss of 30 mass units from m/z 138) allows selectively not unlike that afforded by radiolabels.

Through the application of desorption ionization LC/MS and tandem MS technologies, the determination and identification of polar metabolites and metabolic conjugates may finally become routine, 160 years after the work of von Liebig!

Acknowledgments

The work was performed in part at the California Department of Health Services Hazardous Materials Laboratory. This study was supported in part by the Superfund Program Project No. 04705 from the National Institute of Environmental Health Sciences of the National Institutes of Health.

Literature Cited

1. Doull, J.; Klassen, C.; Amdur, M., Eds. Toxicology, The Basic Science of Poisons; Academic Press: New York, 1966.
2. Caldwell, J. In Xenobiotic Conjugation Chemistry; Paulson, G.; Caldwell, J.; Hutson, D.; Menn, J., Eds.; ACS Symposium Series No. 299; American Chemical Society: Washington, DC 1986; pp 2-28.
3. Hiron, P. C.; Millburn, P. In Foreign Compound Metabolism in Mammals, Volume 5; Hathway, D. E., Ed.; The Chemical Society: London, 1979; pp 132-158.
4. Paulson, G.; Caldwell, J.; Hutson, D.; Menn, J., Eds. Xenobiotic Conjugation Chemistry; ACS Symposium Series No. 299; American Chemical Society: Washington, DC, 1986.
5. Crawford, M. J.; Hutson, D. H. In Bound and Conjugated Pesticide Residues; Kaufman, D.; Still, G.; Paulson, G.; Bandal, S., Eds.; ACS Symposium Series No. 29; American Chemical Society: Washington, DC, 1979; pp 181-229.
6. Snyder, R.; Parke, D. V.; Kocsis, J. J.; Jollow, D. J.; Gibson, C. G.; Witmer, C. M., Eds. Biological Reactive Intermediates; Plenum Press: New York, 1982.
7. Bakke, J. E. In Bound and Conjugated Pesticide Residues; Kaufman, D.; Still, G.; Paulson, G.; Bandal, S., Eds.; ACS Symposium Series No. 29; American Chemical Society: Washington, DC, 1979; pp 55-67.
8. Paulson, G.; Simpson, M.; Giddings, J.; Bakke, J.; Stolzenberg, . Biomed. Mass Spectrom. 1978, 5, 413.
9. Fenselau, C.; Yellet, L. In Xenobiotic Conjugation Chemistry; Paulson, G.; Caldwell, J.; Hutson, D.; Menn, J., Eds.; ACS Symposium Series No. 299; American Chemical Society: Washington, DC, 1986; pp 159-176.
10. Liberato, D. J.; Fenselau, C. C.; Vestal, M. L.; Yergey, A. L. Anal. Chem. 1983, 55, 1741.
11. Watson, D.; Taylor, G. W.; Murray, S. Biomed. Environ. Mass Spectrom. 1986, 13, 65.
12. Betowski, L. D.; Korfmacher, W. A.; Lay, J. O., Jr.; Potter, D. W.; Hinson, J. A. Biomed. Environ. Mass Spectrom. 1987, 14, 705.
13. Blake, T. J. A. J. of Chromatogr. 1987, 394, 171.
14. Korfmacher, W. A.; Holder, C. L.; Betowski, L. D.; Mitchum, R. K. J. Anal. Toxicol. 1987, 11, 182.

15.   Voyksner, R. D.; Hagler, W. M., Jr.; Swanson, . J. Chromatogr. 1987, 394, 183.
16.   Draper, W. M.; Brown, F. R.; Bethem, R.; Miille, M. J. Biomed. Environ. Mass Spectrom. 1989, 18, 767.
17.   Watson, D.; Taylor, G. W.; Murray, S. Biomed. Environm. Mass Spectrom. 1985, 12, 610.
18.   Liberato, D. J.; Yergey, A. L. Anal. Chem. 1986, 58, 6.
19.   Brown, F.R.; Draper, W.M. J. Chromatogr. 1989,479,441-444.
20.   Voyksner, R.D.; Bursey, J.T.; Pellizari, E.D. Anal. Chem. 1984, 56, 1507.
21.   Blakeley, C.R.; Vestal, M.L. Anal. Chem. 1983, 55, 750.
22.   Bursey, M. M.; Parker, C. E.; Smith, R. W.; Gaskell, S. J. Anal. Chem. 1985, 57, 2597.
23.   Voyksner, R. D.; Haney, C. A. Anal. Chem. 1985, 57, 991.
24.   Barcelo, D. Biomed. Environ. Mass Spectrom. 1988, 17, 363.
25.   Voyksner, R. D.; Yinon, J. J. Chromatogr. 1986, 354, 393.
26.   Vajta, S.; Thenot, J. P.; De Maack, F.; Devant, G.; Lesieur, M. Biomed. Environ. Mass Spectrom. 1988, 15, 223.
27.   Lee, M. S.; Yost, R. A. Biomed. Environ. Mass Spectrom. 1988, 15, 193.
28.   Straub, K. M.; Rudewicz, P.; Garvie, C. Xenobiotica 1987, 17, 413.

RECEIVED October 16, 1989

Chapter 18

Structural Studies of In Vitro Alkylation of Hemoglobin by Electrophilic Metabolites

S. Kaur[1], D. Hollander[2], R. Haas[2], and A. L. Burlingame[1]

[1]Mass Spectrometry Facility, Department of Pharmaceutical Chemistry, School of Pharmacy, University of California, San Francisco, CA 94143–0446
[2]Air and Industrial Hygiene Laboratory, California Department of Health Services, 2151 Berkeley Way, Berkeley, CA 94704

Alkylation of DNA by xenobiotic agents, or their electrophilic metabolites, is believed to be the major initiating process that may result ultimately in carcinogenesis. The study of hemoglobin alkylated *in vivo* by chemical carcinogens has previously been proposed as an indicator of DNA alkylation. Xenobiotically modified proteins, however, are not readily amenable to conventional methods for amino acid sequencing. Tandem mass spectrometry allows unambiguous structural elucidation of chemically modified proteins. Styrene is a widely used chemical in the plastics industry and its major metabolite, styrene-7,8-oxide, is both mutagenic and carcinogenic in rodents. Human hemoglobin was modified *in vitro* with styrene-7,8-oxide and digested with trypsin. Tryptic peptides from unmodified hemoglobin were isolated by HPLC and their molecular weights were determined by liquid secondary ion mass spectrometry. This allowed confirmation of the known sequence of the protein and provided a reference for the identification of modified peptides. High performance tandem mass spectrometry of modified peptides allowed unambiguous assignment of specific residues modified at the low pmol level. The externally accessible histidines were found to be the dominant sites for alkylation at high modification levels of the protein.

0097–6156/90/0420–0270$06.00/0

Epidemiological evidence suggests that many industrial chemicals are carcinogenic. Although the parent compound itself may be inactive, metabolism *in vivo* can produce species capable of attacking nucleophilic sites in DNA resulting in covalent adducts (1-2). This process is believed to play a major role in the initiation of chemical carcinogenesis (1-2). Electrophilic metabolites formed may also attack nucleophilic sites in a variety of proteins (3). It has been suggested that the modification of hemoglobin *in vivo* by carcinogens may be an indicator of DNA alkylation (4). The structural assignment of hemoglobin adducts may therefore provide a ready means of identifying metabolites of chemical carcinogens and monitoring exposure for the purposes of risk assessment. In contrast to DNA alkylation, alkylation in hemoglobin is not repaired (5). This, together with the long biological lifetime of the protein (18 weeks in humans), allows the assessment of long-term occupational exposure (6). Covalent DNA adducts from target cells are generally difficult to obtain in sufficient quantity for structural studies, particularly since modifications may be removed by normal repair processes. The abundance of hemoglobin in blood (2000 nmol per ml) point to the feasibility of structural studies.

Previous studies of hemoglobin modified by electrophilic agents have included cleavage of the N-terminal valine adducts by reaction with pentafluorophenylisothiocyanate and analysis by GC-MS (7). This approach is limited to the identification of N-terminal modification which generally represents only a small fraction of the total covalent modification of hemoglobin (e.g. 3% for styrene oxide). Other approaches involve basic hydrolysis of the protein followed by derivatization and GC-MS analysis (8,9) or non-specific digestion followed by FAB-MS (10,11). The latter approaches facilitate analysis of the entire suite of adducts formed, however, specific amino acid residues modified are not likely to be revealed. In contrast, recently developed tandem mass spectrometric (tandem MS) techniques offer the potential for the unambiguous structural determination of the suite of modified peptides including the specific site(s) of modification (12). Our strategy has involved the determination of molecular masses of tryptic peptides from native human hemoglobin by liquid secondary ion mass spectrometry (LSIMS) and using the information as a database for the comparison of covalently modified

hemoglobin. Since modifications are detected as mass
shifts relative to the corresponding unmodified
peptides, the nature of the modifying electrophile need
not be known. Hence, this approach offers the potential
of identifying human exposure to unknown metabolites of
chemical carcinogens. The specific residues modified
have been identified by tandem MS experiments using an
optically-coupled multichannel array detector (13).

   Styrene-7,8-oxide (styrene oxide), the major
metabolite of the commercially important chemical
styrene, was used as a model electrophile. Styrene is
widely used in the manufacture of reinforced plastics
and occupational exposure occurs mainly through
inhalation of the vapor (14). The metabolite styrene
oxide is mutagenic in both prokaryotic (15) and
eukaryotic test systems (16,17) and carcinogenic in
rodents (18). The formation of covalent DNA adducts
with styrene oxide in vitro has been reported (19),
therefore the development of procedures with the
potential for the identification and assessment of
styrene oxide damage in vivo clearly need to be
explored.

Experimental Procedures

Materials. Starting materials were obtained as follows:
Fresh whole blood was obtained as a "short unit" (Irwin
Memorial Blood Bank, San Francisco, CA); hemoglobin was
prepared from washed red blood cells by cell lysis using
three volumes of sterile H_2O at 4°C. The cell debris
was removed by centrifugation (12 000 g, 15 min), and
residual serum proteins were precipitated in 20% (v/v)
ammonium sulfate. The hemoglobin-containing supernatant
was dialysed against 0.01 M Bis-Tris (pH 7.0); [8-^{14}C]
Styrene-oxide 25mCi/mmol, (Amersham Corp.); Styrene
oxide (Aldrich); TPCK-treated trypsin, (Worthington);
All solvents were HPLC grade.

Instrumentation. HPLC isolations were performed on a
Beckman 421A system using a Vydac column (C-18, 4.6 x
250 mm). Liquid secondary ion mass spectra (LSIMS) were
recorded in the positive ion mode on a Kratos
(Manchester, UK) MS-50S mass spectrometer equipped with
a 23 kG magnet and post-acceleration detector. The
LSIMS ion source has been described elsewhere (20). A
Cs^+ ion beam of energy 10 keV was used as the primary
beam (21). Spectra were recorded (300 sec per decade)
with a Gould ES-1000 electrostatic recorder. Tandem MS
experiments were performed on a Kratos Concept IIHH
(Manchester, UK) four sector instrument of EBEB

geometry. The sample was ionized as above in a LSIMS ion source using a Cs^+ beam of energy 12 keV. Only the ^{12}C isotope peak for the MH^+ ion cluster was selected in MS-1 and introduced into a collision cell containing helium. The collision energy was 6 keV and the helium pressure in the cell was adjusted to obtain a MH^+ ion attenuation of 65%. The fragment ions generated in the collision cell were separated in MS-2 and detected in successive 4% mass windows (2 sec exposure per frame) using an optically-coupled 1,024 channel array detector (13). Samples for MS analysis were taken to near dryness under vacuum and redissolved in 0.1% TFA/H_2O. A glycerol/-thioglycerol/0.1M HCl (1:1:trace) matrix (1μl) was applied to the stainless steel probe tip and a small aliquot of the sample was added. The sample probe tip was cooled during the MS analysis (22). ^{14}C Radioactivity measurements were obtained by liquid scintillation counting in a Tm Analytic (Elk Grove Village, IL) instrument.

Hemoglobin modification. (i) Hemoglobin (0.14mM) was incubated with 25mM [8-^{14}C] styrene oxide (25mCi/mmol) in 8.5 mM Bis-Tris (pH 7.0), 15% (v/v) DMSO, 0.2% (v/v) EtOH (total volume 0.5 ml) at 20°C in the dark. Unreacted styrene oxide was removed by dialysis against 5 changes of 10 mM sodium phosphate, 150 mM NaCl (1:8000 v/v, pH 7.4). Globin was isolated as a protein precipitate according to the method of Rossi, Fanelli et al. (23). (ii) The above conditions were scaled-up to a total volume of 20 ml using unlabeled styrene oxide. (iii) A reaction mixture without styrene oxide was prepared as a blank.

TPCK-treated trypsin (5μg) was added to an aliquot of the globin (1.3 mg = 20 nmole hemoglobin) from (i) in 150 μl of 0.05 M NH_4HCO_3 (pH 8.5) and the solution was incubated at 37°C for 6h. A further 5 μg of the enzyme was added after 2h and 4h. The tryptic peptides were isolated by reversed phase C-18 HPLC (solvent A: 0.1% TFA, solvent B: 70% acetonitrile/30% H_2O/0.08% TFA; solvent gradient 0-60% B in 60 min at 1 ml min^{-1}). Fractions were collected at one min intervals and the radioactivity was monitored by liquid scintillation counting (Fig. 1b). An aliquot of globin from (ii) prepared as above (3.5 mg = 54.3 nmole hemoglobin) was taken up in 400 μl of 0.05 M NH_4HCO_3 (pH 8.5) and 40 μg TPCK-treated trypsin was added. The solution was incubated at 37°C for 6h. A further 10 μg of trypsin was added after 2h and 4h. The tryptic peptides were obtained as above (monitoring at 215 nm), pooling fractions from 5 HPLC analyses (Fig. 1a). LSIMS and CID spectra of selected components were recorded. Tryptic

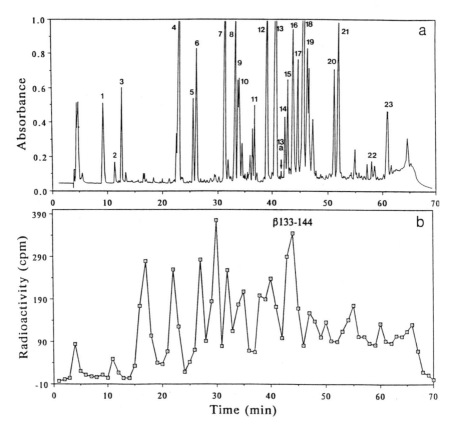

Fig. 1. HPLC of tryptic digest of human globins (α
and β) from hemoglobin treated *in vitro* with [8-^{14}C]
styrene-7,8-oxide (a) chromatogram; numbers refer to
fractions, (b) radiogram; annotation refers to tryptic
peptide.

peptides from globin (0.5mg=10 nmol hemoglobin) from
(iii), the blank experiment, were similarly obtained and
their molecular masses determined by LSIMS. The C-
terminal dipeptides were identified as the hexyl ester
derivatives. The procedure for derivative formation has
been described elsewhere (24).

Results and Discussion

Hemoglobin was isolated from human erythrocytes and
globin was prepared by removal of the heme moiety. The
globin (α and β chains) was enzymatically digested with
trypsin. The resulting tryptic peptides were fractionated by HPLC (*cf.* Fig. 1a) and the molecular weights
were determined by LSIMS (Table 1). The amino acid
sequences for human α and β globins are shown in Figure
2. The tryptic peptides isolated were identified by
comparison of the molecular weights obtained by LSIMS
with those calculated for the known sequences (Table 1,
Fig. 2). Molecular weights were obtained for all the
fractions indicated in Figure 1 with the exception of
fractions 1 and 3. Low molecular weight hydrophilic
peptides are not generally detected by LSIMS (25).
However, the introduction of hydrophobicity in such
peptides by derivative formation may lead to significant
improvements in the LSIMS spectrum (25). A recently
reported single step procedure for the formation of
hexyl ester derivatives was used to obtain derivatives
for peptides in fractions 1 and 3 (24). LSIMS spectra
after derivatization showed molecular ions MH^+ m/z 422.2
and m/z 403.1 for fractions 1 and 3, respectively.
This identified the α and β C-terminal dipeptides Tyr-
Arg and Tyr-His, respectively (Fig. 3, Table 1). LSIMS
confirmed 97% of the amino acid sequence for α globin
and 96% of the sequence for β globin. Tryptic peptides
not detected represented low molecular weight hydrophilic peptides: Gly-His-Gly-Lys, α(57-60); Val-Lys,
β(60-61); Ala-His-Gly-Lys, β(62-65). It is likely that
these peptides were not retained under the
chromatographic conditions used.
 Hemoglobin isolated from human erythrocytes was
covalently modified by incubating with (i) $[^{14}C-8]$
styrene oxide and (ii) unlabeled styrene oxide. Excess
styrene oxide was removed by dialysis and the level of
covalent binding was determined by liquid scintillation
counting (*ca.* 1% w/w). Tryptic peptides were obtained
as above and fractionated by HPLC, monitoring by liquid
scintillation counting. The radiogram indicated a
number of peptides containing covalently bound styrene
oxide (Fig. 1b). The HPLC retention behavior of the
radiolabeled covalent adducts from (i) was used to
locate the corresponding adducts from the reaction of

α-Chain

```
V L S P A D K T N V K A A W G K V G A H A G E Y G A E A L E
1             8         12            17
R M F L S F P T T K T Y F P H F D L S H G S A Q V K G H G K
  32                    41                            57
K V A D A L T N A V A H V D D M P N A L S A L S D L H A H K
  62
L R V D P V N F K L L S H C L L V T L A A H L P A E F T P A
91  93          100
V H A S L D K F L A S V S T V L T S K Y R
              128                   140
```

β-Chain

```
V H L T P E E K S A V T A L W G K V N V D E V G G E A L G R
1               9               18
L L V V Y P W T Q R F F E S F G D L S T P D A V M G N P K V
31                    41                              60
K A H G K K V L G A F S D G L A H L D N L K G T F A T L S E
  62        67                            83
L H C D K L H V D P E N F R L L G N V L V C V L A H H F G K
          96                  105
E F T P P V Q A A Y Q K V V A G V A N A L A H K Y H
121                   133                     145
```

Fig. 2. Amino acid sequences for human globins (α and β) .

Table 1. Tryptic peptides from human globins (α and β) identified by liquid secondary ion mass spectrometry after C-18 reversed phase HPLC fractionation

Fraction No.	Molecular Ion(s)[a] (MH+)	Peptide
1	403.1[b] (319.1)	β(145-146)
2	461.3 (461.2)	α(8-11)
3	422.2[b] (338.2)	α(140-141)
4	729.4; 531.8 (729.4; 532.3)	α(1-7); α(12-16)
5	1171.6 (1171.7)	α(1-11)
6	952.4 (952.5)	β(1-8)
7	1529.4; 818.6 (1529.7; 818.4)	α(17-31); α(93-99)
8	1378.4 (1378.7)	β(121-132)
9	1149.8 (1149.7)	β(133-144)
10	1314.9 (1314.7)	β(18-30)
11	1087.9; 1449.9 (1087.6; 1449.8)	α(91-99); β(133-146)
12	932.8; 1421.6 (932.5; 1421.7)	β(9-17); β(83-95)
13	1833.8 (1833.8)	α(41-56)
14	1252.2 (1252.7)	α(128-139)
15	1127.0 (1126.6)	β(96-104)
16	1071.6 (1071.6)	α(32-40)
17	1797.5; 2529.0 (1798.0; 2529.2)	β(66-82); β(83-104)
18	1274.8; 2059.0 (1274.7; 2058.9)	β(31-40); β(41-59)
19	1669.5 (1669.9)	β(67-82)
20	3124.4 (3124.6)	α(61-90)
21	2995.9 (2996.5)	α(62-90)
22	2970.1 (2967.6)	α(100-127)
23	1719.8 (1720.0)	β(105-120)

a MH+ observed (MH+ predicted).
b As the hexyl ester derivative.

hemoglobin with unlabeled styrene oxide. The
appropriate fractions were collected and analyzed by
LSIMS to obtain molecular weights of the modified
peptides. For example, modified peptide(s) with
retention time (t_R) 43-44 min in the radioactivity
profile, corresponded to fraction 13a in the HPLC
chromatogram. Subsequent LSIMS of fraction 13a showed
molecular ion MH^+ m/z 1269.6 as the major component
(Fig. 4). The minor components MH^+ m/z 1569.6 and m/z
1833.4 are consistent with peptide β(133-146) containing
one hydroxyphenylethyl modification and peptide α(41-
56), respectively. Peptide β(133-146) would arise from
incomplete digestion at lysine β(144). Peptide α(41-56)
presumably arises from tailing of fraction 13 (Fig.1,
Table 1). The molecular ion MH^+ m/z 1269.6 is
consistent with the peptide 133-144 in the β-chain
containing one hydroxyphenylethyl moiety, indicating
covalent binding of one styrene oxide molecule. The
corresponding unmodified peptide had previously been
identified in fraction 9 (t_R 34.5 min). The increase in
t_R for the modified peptide (t_R 43.5 vs. 34.5 min) is
consistent with an increase in hydrophobicity associated
with aralkylation of the peptide (26).

The specific site of modification was established
by high performance tandem MS experiments. In this
technique the ^{12}C isotope peak for the molecular ion in
the first mass spectrometer is selectively introduced
into an inert gas collision cell where fragmentation is
induced. The fragments are separated and detected in
the second mass spectrometer (27). Collisionally
induced dissociation (CID) provides abundant structural
information from which the amino acid sequence for the
peptide (28) or, in this case, modified peptide can be
deduced.

The CID spectrum (Fig. 5) of fraction 9 (Fig. 1)
confirmed the sequence of the unmodified β(133-144) peptide as Val-Val-Ala-Gly-Val-Ala-Asn-Ala-Leu-Ala-His-Lys.
Significant ions arising from fragmentation of the peptide backbone are indicated in standard nomenclature
(28). For example, the mass difference 99 between the
ions at m/z 1118 (w_{12}) and 1019 (w_{11}) shows the presence
of valine. Similarly, the mass difference 71 between
m/z 355 (y_3) and 284 (y_2) indicates the presence of
alanine at that position. Peaks in the low mass region
of CID spectra represent the immonium ions H_2N^+CHR
(where R is a side-chain group) arising from individual
amino acid residues. The major ion m/z 110 in the
immonium ion region of the spectrum for fraction 9
arises from histidine.

The CID spectrum of the modified β(133-144)
(fraction 13a, 20 pmol globin) is shown in Fig. 6. The

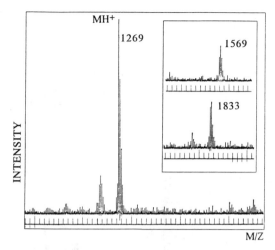

Fig. 3. Liquid secondary ion mass spectrum of fraction 13a showing molecular ion MH+ *m/z* 1269 as the major component, peptide β(133-144). The inset shows the minor components MH+ *m/z* 1569 and *m/z* 1833 consistent with peptide β(133-146) containing one hydroxyphenylethyl modification and peptide α(41-56), respectively. Peptide β(133-146) would arise from incomplete cleavage at Lys β(144).

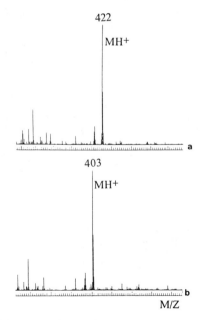

Fig. 4. Liquid secondary ion mass spectra for the C-terminal peptides (a) Tyr-Arg, α(140-141) and (b) Tyr-His, β(145-146) as the hexyl ester derivatives.

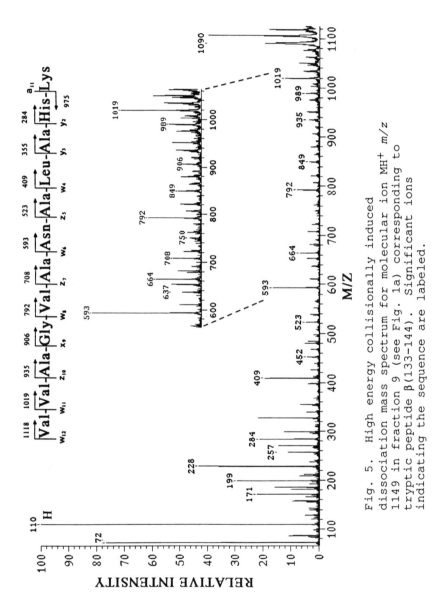

Fig. 5. High energy collisionally induced dissociation mass spectrum for molecular ion MH$^+$ m/z 1149 in fraction 9 (see Fig. 1a) corresponding to tryptic peptide β(133–144). Significant ions indicating the sequence are labeled.

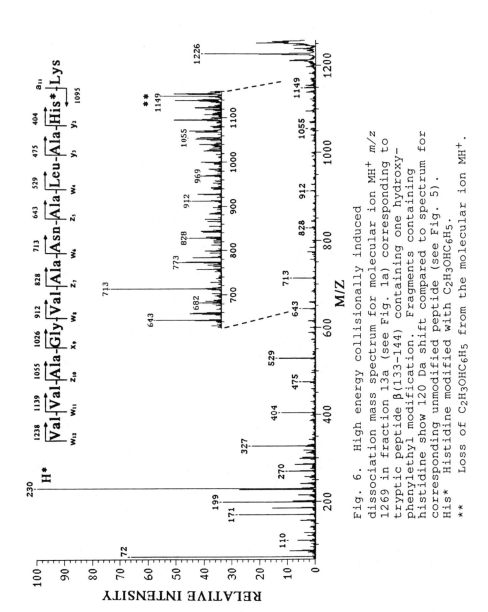

Fig. 6. High energy collisionally induced dissociation mass spectrum for molecular ion MH⁺ m/z 1269 in fraction 13a (see Fig. 1a) corresponding to tryptic peptide β(133–144) containing one hydroxy-phenylethyl modification. Fragments containing histidine show 120 Da shift compared to spectrum for corresponding unmodified peptide (see Fig. 5).
His* Histidine modified with $C_2H_3OHC_6H_5$.
** Loss of $C_2H_3OHC_6H_5$ from the molecular ion MH⁺.

dramatic shift of the histidine immonium ion m/z 110 to m/z 230 clearly identifies the site of modification by way of the 120 Da increase expected for a hydroxyphenylethyl moiety. This is confirmed by fragment ions of the w, x, y and z series containing histidine which show an analogous shift in 120 Da in the modified peptide (Figs. 5 and 6). As expected, ions of the a, b, c and d series (charge retention at the N-terminus) do not show a mass shift except for fragments containing the histidine, e.g. m/z 1095 (a_{11}) compared with m/z 975 (a_{11}) in the unmodified peptide. This again confirms histidine as the specific residue modified. The reactivity of histidine toward attack by electrophiles has previously been indicated by the reaction of polyhistidine with styrene oxide (29). Reactivity of histidine residues *in vivo* has previously been demonstrated by the GC-MS detection of N^{τ}-(2-hydroxyethyl) histidine, as the methyl ester heptafluorobutyryl derivative, following basic hydrolysis of globin samples isolated from workers exposed to ethylene oxide (6).

Due to the low level of modification and the complexity of the peptide mixture, in some cases fractions containing modified peptides were purified by a second stage of HPLC using a phenyl column. Figure 7 illustrates the significant improvement in signal:noise obtained for MH^+ 2073.9 corresponding to peptide α(41-56) containing two hydroxy-phenylethyl modifications. The inset shows the LSIMS spectrum after C-18 fractionation where the signal is barely distinguished from the noise. After further purification over a phenyl column the signal:noise is *ca.* 15:1. Our study identified a number of sites of hemoglobin modification with styrene oxide, including the nucleophilic cysteine β(93), and these results will be published in detail elsewhere.

The application of high performance tandem MS to identify aralkylation of hemoglobin with styrene oxide illustrates the potential of this approach for *in vivo* biological monitoring. At present, however, the strategy outlined is not readily amenable for monitoring occupational exposure in humans. This is largely due to the high levels of unmodified peptides which co-elute during the HPLC fractionation and render difficult the identification of modified peptides. Isolation of modified hemoglobin prior to enzymatic digestion using, for example, immunoaffinity chromatography may result in improvements in this area. The combination of mass spectrometry and immunoassay technology therefore offers exceptional promise for the structural study of carcinogen interactions with macromolecules *in vivo*. The

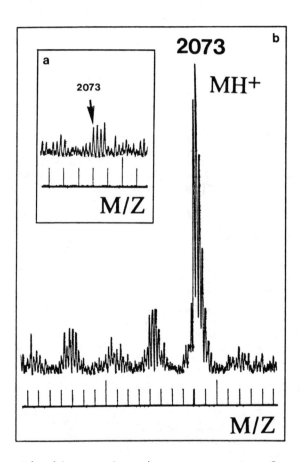

Fig. 7. Liquid secondary ion mass spectra for peptide α(41-56) containing two hydroxyphenylethyl modifications after (a) C-18 reversed phase HPLC and (b) C-18 followed by phenyl reversed phase HPLC.

molecular masses of tryptic peptides from modified
hemoglobin would allow ready identification of covalent
adducts on the basis of mass shifts and HPLC retention
behavior compared with unmodified peptides. Subsequent
high performance tandem MS experiments with multichannel
array detection would allow structural assignment of the
xenobiotic electrophile and the site(s) of substitution
for samples at the pmol level.

Acknowledgments

We gratefully acknowledge the help of Mr. F.C. Walls
with the mass spectral analysis. Supported by NIEHS
grant ES04705, NIH grant RR01614 and NSF grant DIR
8700766.

Literature Cited

1. Miller, J. A.; Miller, E. C. In Origins of Human
   Cancer; Hiatt, H. H.; Watson, J. D.; Winsten J.
   A., Eds.; Cold Spring Harbor Laboratory, 1977; pp
   605-627.
2. Grover, P. L. In Chemical Carcinogens and DNA; CRC
   Press: Florida, 1979.
3. Burlingame, A. L.; Straub, K.; Baillie, T. A.
   Mass Spectrom. Revs. 1983, 2, 331-387.
4. Osterman-Golkar, S., Ehrenberg, L., Segerback, D.
   and Hallstrom, L. Mutat. Res. 1976, 34, 110.
5. Calleman, C. J.; Ehrenberg, L.; Jansson, B.;
   Osterman-Golkar, S.; Segerback, D.; Svensson, K.;
   Wachtmeister, C. A. J. Environ. Pathol. Toxicol.
   1978, 2, 427-442.
6. Farmer, P. B.; Bailey, E.; Gorf, S. M.; Tornqvist,
   M.; Osterman-Golkar, S.; Kautiainen; Lewis-
   Enright, D. P. Carcinogenesis 1986, 7, 637-640.
7. Törnqvist, M.; Mowrer, J.; Jensen, S.; Ehrenberg,
   L. Anal. Biochem. 1986, 154, 255-266.
8. Bailey, E.; Farmer, P. B.; Bird, I.; Lamb, J. H.;
   Peal, J. A. Anal. Biochem. 1986, 157, 241-248.
9. Stillwell, W. G.; Bryant, M. S.; Wishnok, J. S.
   Biomed. Environ. Mass Spectrom. 1987, 14, 221-223.
10. Skipper, P. L.; Obiedzinski, M. W.; Tannenbaum, S.
    R.; Miller, D. W.; Mitchum, R. K.; Kadlubar, F. F.
    Cancer Res. 1985, 45, 5122-5127.
11. Sabbioni, G.; Skipper, P. L.; Buchi, G.;
    Tannenbaum,
    S. R. Carcinogenesis 1987, 8, 819-824.
12. Hutchins, D. A.; Skipper, P. L.; Naylor, S.;
    Tannenbaum, S. R. Cancer Res. 1988, 48, 4756-
    4761.

13. Burlingame, A. L. Proc. 36th Ann. Conf. Mass Spectrom. and Allied Topics, 1988, pp. 727-728.
14. Styrene: Environmental Health Criteria 26, World Health Organization: Geneva, 1983.
15. de Meester, C.; Poncelet, F.; Roberfroid, M.; Rondelet, J.; Mercier, M. Mutat. Res. 1977, 56, 147-152.
16. Bonatti, S.; Abbondandolo, A.; Corti, G.; Fiorio, R.; Mazzaccaro, A. Mutat. Res. 1978, 52, 295-300.
17. Ponomarkov, V.; Cabral, J. R. P.; Wahrendorf, J.; Galendo, D. Cancer Lett. 1984, 24, 95-101.
18. Lijinsky, W. J. Natl. Cancer Inst. 1986, 77, 471-476.
19. Savela, K.; Hesso, A.; Hemminki, K. Chem. Biol. Interactions 1986, 60, 235-246.
20. Falick, A. M.; Wang, G. H.; Walls, F. C. Anal. Chem. 1986, 58, 1308-1311.
21. Aberth, W.; Straub, K. M.; Burlingame, A. L. Anal. Chem. 1982, 54, 2020-2034.
22. Falick, A. M.; Walls, F. C.; Laine, R. A. Anal. Biochem. 1986, 159, 132-137.
23. Rossi Fanelli, A.; Antonini, E.; Caputo, A. (1958), Biochim. Biophys. Acta 1958, 30, 608.
24. Falick, A. M.; Maltby, D. A. Anal. Biochem., in press.
25. Naylor, S.; Findeis, A. F.; Gibson, B.W.; Williams, D. H. J. Am. Chem. Soc. 1986, 108, 6359-6363.
26. Sasagawa, T.; Okuyama, T.; Teller, D. C. J. Chromatogr. 1982, 240, 320-340.
27. Tandem Mass Spectrometry, McLafferty, F. W., Ed.; Wiley: New York, 1983.
28. Johnson, R. S.; Martin, S. A.; Biemann, K. Int. J. Mass Spectrom. Ion Proc. 1988, 86, 137-154.
29. Hemminki, K. Carcinogenesis 1983, 4, 1-3.

RECEIVED December 5, 1989

INDEXES

Author Index

Affiliation Index

Subject Index

Production: Beth Ann Pratt-Dewey
Indexing: Deborah H. Steiner
Acquisition: Cheryl Shanks

Elements typeset by Hot Type Ltd., Washington, DC
Printed and bound by Maple Press, York, PA

Paper meets minimum requirements of American National Standard
for Information Sciences—Permanence of Paper for Printed Library
Materials, ANSI Z39.48–1984 ∞

Other ACS Books